CITY LIMITS

A SOCIAL HISTORY OF QUEENS

Janet E. Lieberman
and
Richard K. Lieberman

This book is based on the work of the Community History Program at Fiorello H. LaGuardia Community College of the City University of New York. The research was funded by the National Endowment for the Humanities.

KENDALL/HUNT
PUBLISHING COMPANY
Dubuque, Iowa

Program Directors:	Janet E. Lieberman Richard K. Lieberman
Research:	Jeffrey Kroessler Vincent Seyfried Linda Ocasio
Literature:	Sarah Barber Harvey Wiener
Ethics:	John Chaffee
Administration	Eduvina Estrella Karl Melder
Design:	Orlie Kraus
Consultants:	Betsy Blackmar Warren Leon Roy Rosenzweig

The authors would like to thank the Board of Directors of the Institute for Neighborhood Research, Inc. for their support and cooperation.

The authors gratefully acknowledge permission to reprint articles and illustrations from the following individuals and organizations:

page 5—Oil painting, courtesy of Dora Geipel.
page 6—Midwife certificate courtesy of Dora Geipel.
page 8—Photo, courtesy of Dora Geipel.
page 14—Photo, courtesy of Dora Geipel.
page 16—From "Return of the Midwife" as seen in Attenzione Magazine.
page 17—From WE WERE THERE: THE STORY OF WORKING WOMEN IN AMERICA by Barbara Wertheimer. Copyright 1977 by Pantheon Books, a Division of Random House, Inc.
page 22—Photo, courtesy of Dora Geipel.
page 24—Photo, courtesy of Dora Geipel.
page 25—Reprinted by permission of Farrar, Strauss and Giroux, Inc. Excerpt from THE ASSISTANT by Bernard Malamud. Copyright ©1957 by Bernard Malamud.
Continued on p. 171.

Cover photo: Courtesy Long Island Division, Queens Borough Public Library.

This edition has been printed directly from the authors' manuscript copy.

Library of Congress Catalog Card Number: 83-82248

ISBN 0-8403-3150-9

Printed in the United States of America
10 9 8 7 6 5 4 3 2

TABLE OF CONTENTS

FOREWORD

History is the oldest of the social sciences, for in the beginning it contained them all. And to a great degree it still does. But what especially distinguishes history from the other social sciences is that for history to be able to do its work it must remain in contact with the ordinary citizen. Economists and sociologists can speak to one another in their own jargon, which no one outside their professions understands. But doing that does not really affect the citizen because what we and the citizen expect from those social scientists is not communication or even understanding, but the solution to economic and social problems. History, though, is less concerned with solving problems and more with helping us to understand how we got to where we are today and who we are. It is, fundamentally, collective memory. That is why everyone wants and needs history; without it, as without a memory, we have no sense of where we have come from or who we are. In order for people to know their past, as individuals or as members of a group or of a nation, that history must be understandable, accessible to the ordinary citizen. That is why history must be written and taught in the language of people, not in the jargon of professionals or scholars remote from every day life.

It is also important that ordinary people have a direct acquaintance with how history is compiled and written. In that way they see it as something everyone does with his or her own life. Every personal document, no matter how trivial, is a historical document, for it tells about a human being's past. And that is why this book, *City Limits,* which Dr. Janet Lieberman and Dr. Richard Lieberman have put together, is so important. These documents make history both a means of learning and a means of understanding who one is by bringing to students documents about the neighborhood in which they live. Suddenly, in these documents, students see that history is not something abstract or remote from their everyday experience, as national history too often appears to them, but that it is all around them. These documents are more than just relics of a local region. Again and again the compilers of these documents have pointed out how what is learned about Astoria can be related to larger, often national themes, such as the writings of the great American novelist, F. Scott Fitzgerald, or the relation between work in the nineteenth century and a job today. The work records of the Steinway Company in the 1880s will throw a flood of light for students on the practices of business and the state of workers in the nineteenth century. From the quality of the penmanship of the entries on the work records to the reasons for firing a worker, these first-hand documents provide an exciting opportunity for a student to become his or her own historian, drawing conclusions and making interpretations from original sources. Perhaps for the first time, students using these records can do what historians do professionally: put the disorganized, chaotic record of the past into some order by letting their imaginations as well as their logical analyses play over documents written by people of another time for another purpose and with another interest. Again, perhaps for the first time, students will stand face to face with the past without a historian standing between them. At that point students will recognize that within their own family, and within their own neighborhood, that kind of original contact with the past also exists in the form of letters, photos, perhaps diaries, or business records left by people who no longer are alive. It will then become clear that every person is, in the end, his or her own historian. For, through history, men and women learn not only about the past, but about themselves and the present as well.

Carl N. Degler
Margaret Byrne Professor of
American History
Stanford University

INTRODUCTION

> "To mistake maps for landscapes is to degrade every other way of knowing the world's terrain.... We forget that to map a forest means less than to write poems about it; to map a village means less than to visit among its people; to map a sacred grove means less than to worship there. Making maps may be absorbing and useful; it may take enormous intellectual talent and great training; but it is the most marginal way of knowing the landscape."

To study the history of Queens is to visit among its people. In this book you will explore the landscapes and lineage of Queens through documents and literature. The documents and related literature are organized in three separate themes: **Work, Leisure and Transportation.**

While these documents can be read for information about **Work, Leisure and Transportation**, we hope you will also find in them some information about the nature of history itself. Reading the narrative and analyzing the documents show that the past is not simply about the famous, rich, or powerful. History is not a series of facts to be memorized. Rather, history is interpreting the documents left behind by ordinary people.

Recently, people have shown a growing interest in the social history of their community. Unfortunately, it is difficult to find the documents which will reveal the texture of everyday life. This book offers you a unique chance to look closely at ordinary life in Queens during the nineteenth and twentieth centuries. The documents reproduced in the following pages come largely from personal collections of people living in Queens.

We invite you to become an historian rather than merely a consumer of facts. Look at the pictures, read the narrative, study the questions and think about the past—when it was different. Then investigate the rich and complex story of your own neighborhood. We hope you will use this book as a starting point for locating and interpreting documents, objects and recollections about your own history.

WORK

INVITATION TO SEEING IN PERSPECTIVE

Today, Americans try to bring personal meanings to work. We work for economic reasons, to pay for the necessities of life or for its luxuries. We work for social reasons; work may be where we make friends. In a world where extended families, neighborhoods and churches are becoming less important sources of community life, people look forward to socializing with their co-workers. We also work for psychological reasons. Workers often find themselves, their identities in their jobs, saying "I am a nurse," or "I am a carpenter." Many people get a sense of satisfaction and self-worth from their work. Some people look for security; others look for themselves in their work. Some work to support their families; others work to escape unpleasant domestic conditions.

Obviously, we each have values, needs, and desires that affect our attitudes toward work. The material in this section shows how people from different walks of life coped with the question of work at different times. It shows the interaction of their work and their lives, and also suggests how artists, poets, and writers have expressed the emotions and attitudes connected to work and experienced by workers. Our appreciation is increased when we read and look at the artists' and writers' interpretation. In addition, learning about the history of work and its changing patterns helps us understand what's happening to us today.

The story of work, family, and community in Astoria and Long Island City in the past 100 years is the story of families like the Geipels, the Bonner-Cannon-Aarons, and the Manyokys—families you will meet in this section. Through their experience, you can trace some of the basic changes which have affected most families—including your own—in Queens and elsewhere over the past century.

In addition, by looking at the experiences of these three families, you can see that work is an integral, not a separate part of life. This section includes both typical and unusual family stories. It also suggests how complex real lives are, and how families and communities fit into the major themes of history.

The work of Louise Wilhelmine Geipel and August Geipel, as midwife and saloonkeeper, could only have taken place within the German immigrant community which flourished in Astoria around 1900. Although the work of the Bonner-Cannon-Aaron family was similarly shaped by a Queens Black community, the family itself was an even more important force in job choice and lifestyle. Finally, for Emory and Jess Manyoky, it was a particular employer, the Steinway Piano Company, that defined their work world for a total of 90 years between them.

THE GEIPEL FAMILY

Louise Wilhelmine Geipel and August Frederick Geipel emigrated from Germany to Astoria, Queens, in the 1870's. Wilhelmine worked as a midwife from 1875 to 1904, and August was a brewmaster for the George Ehret Brewery until 1880 when he opened his own saloon on Newton Road in Astoria.

This rich collection of documents about the Geipel family has fascinating implications. Here, in 1880, is a family where both the husband and the wife worked. Although Mrs. Geipel may have been unusual in her time, it is clear that women, even middle-class women, worked then. Not only did she work as a midwife, but she owned property and was a major figure in the community.

The data also show that for August Geipel work came to mean free enterprise; he went into business for himself and ran it for profit. The Geipels, like many couples today, had to decide what the relationship would be between their work, their family, and each other.

Through the lives of the Geipels we can begin to understand the world of work in Queens in the late 1800's. This was a world before the development of factories in Queens. It was also a world in which midwives like Mrs. Geipel provided close, personal service to women at home, and saloons like Mr. Geipel's provided a social center for other German immigrants living in the surrounding neighborhood. For the Geipels—as for other residents of Queens and the United States in this era—work, family, community, and ethnicity were all closely related.

OIL PAINTING OF WILHEMINA GEIPEL

Louise Wilhelmine Geipel was an unusually active woman. She worked as a midwife in Astoria from 1875 to 1904, and became an important member of the German community there. Midwives at this time played important roles. A midwife wasn't simply a person who delivered a baby and then disappeared until needed again. She was a friend and a neighbor, a woman who watched the children she brought into the world grow and develop into mature human beings.

Mrs. Geipel received a midwife's certificate in Germany in 1875, and she and her husband came to America a short time later, at a time when many other Germans were also immigrating to this country. German immigration to New York City reached its peak in 1882; by then the Geipels had established themselves in Astoria.

Imagine how complicated Mrs. Geipel's work and life were: she had a professional career as a midwife; she had a responsibility to her home and her children; she was very active in her community, and she probably helped her husband in the saloon as well.

PORTRAIT OF WILHELMINE GEIPEL. *Artist and date unknown*

QUESTIONS: IDEAS TO THINK ABOUT

1. *Judging from this painting, how would you describe Mrs. Geipel's character?*
2. *Why do people have their portraits painted?*

MIDWIFE CERTIFICATE

In the latter part of the nineteenth century women actually preferred having a baby at home with a midwife over having a child in a hospital. The list of births shows that Mrs. Geipel assisted in 55 births in 1881. In the early 1900's, however, midwives began to face stiff competition from the "newly" educated male doctors looking to establish their private practices.

Ironically enough, another reason women stopped being midwives during the early part of the twentieth century was the increasing acceptance of Victorian attitudes about the proper role and place of women in society. It became unthinkable for women to face the facts of medicine, or to mix men and women in professional training, even for such an event as birth. Although it was the women who gave birth, it was not considered "proper" for them to assist in this event.

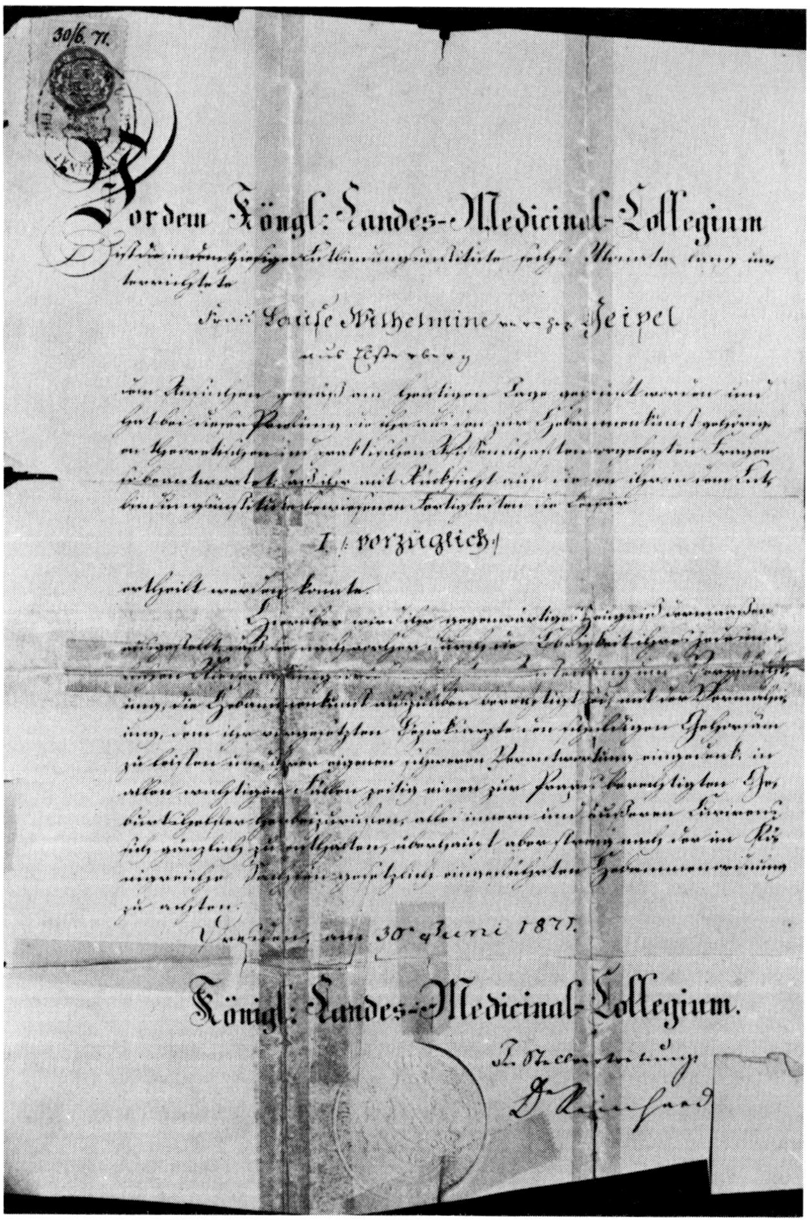
Vor dem Königl: Landes-Medicinal-Collegium

Louise Wilhelmine Geipel

I. vorzüglich

am 30. Juni 1871.

Königl: Landes-Medicinal-Collegium.

MIDWIFE CERTIFICATE, 1871.

The translation of her midwife certificate from 1871 awarded in Dresden, Germany reads as follows:

Imperial Landes School
Medical Institute for Midwives
For Midwife Mrs. Louise Wilhelmine Geipel
From Elsterberg

It is testified with this that she has acquired such a skill-fullness in the help of confinements which she can use in her employment and obligation for sick women but not without assistance from a doctor, who has the responsibiliy . . . The doctors who taught her are very satisfied with her and agree with the institution: She is a very good midwife. She has to follow the laws of the midwife.

Dr. Reinhard - Dresden, June 30, 1871

(Translation from original)

QUESTIONS: IDEAS TO THINK ABOUT

1. *Why do you think the certificate says that the midwife can only practice with "assistance from a doctor who has the responsibility"? What does that imply about the relationship between midwives and male doctors?*
2. *Mrs. Geipel was trained in Germany but practiced in America. Is it possible for professionals to do that today? Why or why not?*
3. *What are the possible reasons why midwifery declined as a career for women? In what ways could a midwife help women, besides delivering babies?*

BIRTHS OF THE YEAR 1881

This document shows the names of the children delivered by Wilhelmine Geipel in 1881. Reading across the page, one finds the date of birth, the name of the child, the sex, the neighborhood (A-Astoria, S-Sunnyside, WO-Woodside, N-Newtown), and the weight of the baby.

A quick look at the names of the families tells us that Mrs. Geipel's practice included mostly German families. Included in this list is a child by the name of Ehret, a new member of the successful brewer's family. The largest group of immigrants coming into New York at that time were Germans. Many of them settled by religion in Catholic, Lutheran, or Jewish areas and by social class in the working class or middle class neighborhoods of the Lower East Side or Yorkville, and moved to the growing Astoria community shortly thereafter.

BIRTHS OF THE YEAR (1881).

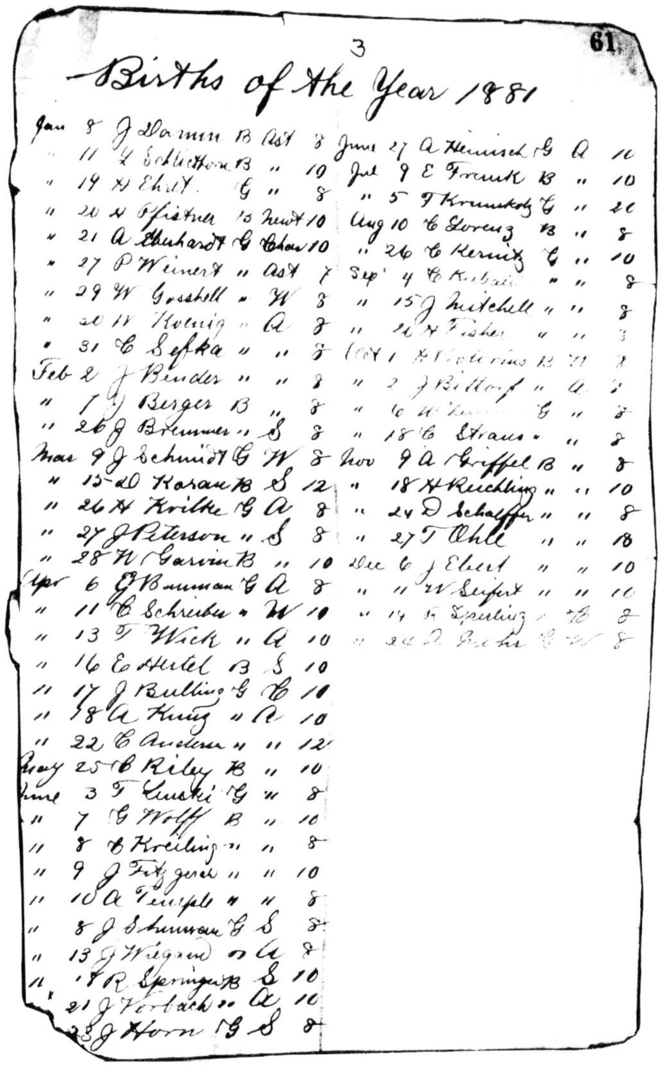
3 61

Births of the Year 1881

Jan	8	J Damm	B	Ast	8
"	11	L Schlichtom	B	"	10
"	19	H Ehret	G	"	8
"	20	H Pfistner	B	Newt	10
"	21	A Eberhardt	G	Chas	10
"	27	P Weinert	"	Ast	7
"	29	W Gosshell	"	W	8
"	30	H Koenig	"	A	8
"	31	C Sefka	"	"	8
Feb	2	J Bender	"	"	8
"	7	J Berger	B	"	8
"	26	J Brenner	"	S	8
Mar	9	J Schmidt	G	W	8
"	15	D Karau	B	S	12
"	26	H Krilke	G	A	8
"	27	J Peterson	"	S	8
"	28	W Garvin	B	"	10
Apr	6	G Baumann	G	A	8
"	11	C Schreiber	"	W	10
"	13	T Wick	"	A	10
"	16	E Hertel	B	S	10
"	17	J Bulling	G	C	10
"	18	A Kunz	"	R	10
"	22	C Andersen	"	"	12
May	25	C Riley	B	"	10
June	3	T Luski	G	"	8
"	7	G Wolff	B	"	10
"	8	C Kreiling	"	"	8
"	9	J Fitzgerald	"	"	10
"	10	A Teufel	"	"	8
"	8	J Schumann	G	S	8
"	13	J Wiegand	"	A	8
"	18	R Springer	B	S	10
"	21	J Forbach	"	A	10
"	23	J Horn	B	S	8
June	27	A Heinisch	G	A	10
Jul	9	E Franck	B	"	10
"	5	T Krumholz	G	"	20
Aug	10	C Lorenz	B	"	8
"	26	C Kernitz	G	"	10
Sep	4	C Kubai[illegible]	"	"	8
"	15	J Mitchell	"	"	8
"	2[illegible]	H Fisher	"	"	8
Oct	1	H Victorius	B	"	8
"	2	J Bittorf	"	A	8
"	6	W [illegible]	G	"	8
"	18	C Straus	"	"	8
Nov	9	A Griffel	B	"	8
"	18	H Reichling	"	"	10
"	24	D Schaffer	"	"	8
"	27	T Ohle	"	"	18
Dec	6	J Ebert	"	"	10
"	11	W Seifert	"	"	10
"	14	[illegible] Sperling	"	B	8
"	24	[illegible]	G	W	8

"... and meantime [the midwife] removed her bonnet and laid it on the mantelpiece, and got out of her bag, first an old dress, and then a saucer of goose-grease, which she proceeded to rub on her hands. The more cases this goosegrease is used in, the better luck it brings to the midwife." —Upton Sinclair, The Jungle

QUESTIONS: IDEAS TO THINK ABOUT

1. *According to the information presented in this document, do you believe that Mrs. Geipel was a successful midwife? Why or why not?*
2. *If she charged $10. for each child she delivered, how much did she earn in 1881?*
3. *How might Mrs. Geipel's earnings have affected her position in her household?*

THE JUNGLE
by Upton Sinclair

To have a baby in the new world and to see it survive is a dream that haunts the immigrant's imagination.

In 1906, Upton Sinclair, a muckraking journalist (a writer who stirs up social issues), wrote *The Jungle,* a novel about the immigrants working in the slaughterhouses in Chicago. The novel tells the story of a Lithuanian immigrant attempting to make a life for his family. One of the central scenes in the book is the moment of his wife's death in childbirth. Because they were poor, they hesitated calling for the midwife until it was too late. Sinclair does not paint a flattering portrait of the midwife, a Dutch woman who expects to be paid even though both mother and child died.

THE JUNGLE, BY UPTON SINCLAIR

"Madame Haupt, Hebamme," ran a sign, swinging from a second-story window over a saloon on the avenue; at a side door was another sign, with a hand pointing up a dingy flight of steps. Jurgis went up them, three at a time.

Madame Haupt was frying pork and onions, and had her door half open to let out the smoke. When he tried to knock upon it, it swung open the rest of the way, and he had a glimpse of her, with a black bottle turned up to her lips. Then he knocked louder, and she started and put it away. She was a Dutch woman, enormously fat—when she walked she rolled like a small boat on the ocean, and the dishes in the cupboard jostled each other. She wore a filthy blue wrapper, and her teeth were black.

"Vot is it?" she said, when she saw Jurgis.

He had run like mad all the way and was so out of breath he could hardly speak. His hair was disordered and his eyes wild—he looked like a man that had risen from the tomb. "My wife!" he panted. "Come quickly!"

Madame Haupt set the frying pan to one side and wiped her hands on her wrapper. "You vant me to come for a case?" she inquired.

"Yes," gasped Jurgis.

"I haf yust come back from a case," she said. "I haf had no time to eat my dinner. Still—if it is so bad—"

"Yes—it is!" cried he.

"Vel, den, perhaps—vot you pay?"

"I—I—how much do you want?" Jurgis stammered.

"Twenty-five dollars."

His face fell. "I can't pay that," he said.

The woman was watching him narrowly. "How much do you pay?" she demanded.

"Must I pay now—right away?"

"Yes; all my customers do."

"I—I haven't much money," Jurgis began in an agony of dread. "I've been in—trouble—and my money is gone. But I'll pay you—every cent—just as soon as I can; I can work—"

"Vot is your work?"

"I have no place now. I must get one. But I—"

"How much haf you got now?"

He could hardly bring himself to reply. When he said, "A dollar and a quarter," the woman laughed in his face.

"I vould not put on my hat for a dollar und a quarter," she said.

"It's all I've got," he pleaded, his voice breaking. "I must get some one—my wife will die. I can't help it—I—"

Madame Haupt had put back her pork and onions on the stove. She turned to him and answered, out of the steam and noise: "Git me ten dollars cash, und so you can pay me the rest next mont'."

"I can't do it—I haven't got it!" Jurgis protested. "I tell you I have only a dollar and a quarter."

The woman turned to her work. "I don't believe you," she said. "Dot is all to try to sheat me. Vot is de reason a big man like you has got only a dollar and a quarter?"

"I've just been in jail," Jurgis cried—he was ready to get down upon his knees to the woman—"and I had no money before, and my family has almost starved."

"Vere is your friends, dot ought to help you?"

"They are all poor," he answered. "They gave me this. I have done everything I can—"

"Haven't you got notting you can sell?"

"I have nothing, I tell you—I have nothing," he cried, frantically.

"Can't you borrow it, den? Don't your store people trust you?" Then, as he shook his head, she went on: "Listen to me—if you git me you vill be glad of it. I vill save your wife und baby for you, und it vill not seem like mooch to you in de end. If you loose dem now how you tink you feel den? Und here is a lady dot knows her business—I could send you to people in dis block, and dey vould tell you—"

Madame Haupt was pointing her cooking fork at Jurgis persuasively; but her words were more than he could bear. He flung up his hands with a gesture of despair and turned and started away. "It's no use," he exclaimed—but suddenly he heard the woman's voice behind him again:—

"I vill make it five dollars for you."

She followed behind him, arguing with him. "You vill be foolish not to take such an offer," she said. "You von't find nobody to go out on a rainy day like dis for less. Vy, I haf never took a case in my life so sheap as dot. I couldn't pay mine room rent—"

Jurgis interrupted her with an oath of rage. "If I haven't got it," he shouted, "how can I pay it? Damn it, I would pay you if I could, but I tell you I haven't got it. I haven't got it! Do you hear me—I haven't got it!"

He turned and started away again. He was halfway down the stairs before Madame Haupt could shout to him: "Vait! I vill go mit you! Come back!"

He went back into the room again.

"It is not goot to tink of anybody suffering," she said, in a melancholy voice. "I might as vell go mit you for notting as vot you offer me, but I vill try to help you. How far is it?"

"Three or four blocks from here."

"Tree or four! Und so I shall get soaked! Gott in Himmel, it ought to be vorth more! Vun dollar und a quarter, und a day like dis! But you understand now—you vill pay me de rest of twenty-five dollars soon?"

"As soon as I can."

"Some time dis mont?"

"Yes, within a month," said poor Jurgis. "Anything! Hurry up!"

"Vere is de dollar und a quarter?" persisted Madame Haupt, relentlessly.

Jurgis put the money on the table and the woman counted it and stowed it away. Then she wiped her greasy hands again and proceeded to get ready, complaining all the time; she was so fat that it was painful for her to move, and she grunted and gasped at every step. She took off her wrapper without even taking the trouble to turn her back to Jurgis, and put on her corsets and dress. Then there was a black bonnet which had to be adjusted carefully, and an umbrella which was mislaid, and a bag full of necessaries which had to be collected from here and there—the man being nearly crazy with anxiety in the meantime. When they were on the street he kept about four paces ahead of her, turning now and then, as if he could hurry her on by the force of his desire. But Madame Haupt could only go so far at a step, and it took all her attention to get the needed breath for that.

They came at last to the house, and to the group of frightened women in the kitchen. It was not over yet, Jurgis learned—he heard Ona crying still; and meantime Madame Haupt removed her bonnet and laid it on the mantelpiece, and got out of her bag, first an old dress and then a saucer of goosegrease, which she proceeded to rub upon her hands. The more cases this goosegrease is used in, the better luck it brings to the midwife, and so she keeps it upon her kitchen mantelpiece or stowed away in a cupboard with her dirty clothes, for months, and sometimes even for years.

Then they escorted her to the ladder, and Jurgis heard her give an exclamation of dismay. "Gott in Himmel, vot for haf you brought me to a place like dis? I could not climb up dot ladder. I could not git troo a trap-door! I vill not try it—vy, I might kill myself already. Vot sort of a place is dot for a woman to bear a child in—up in a garret, mit only a ladder to it? You ought to be ashamed of yourselves!" Jurgis stood in the doorway and listened to her scolding, half drowning out the horrible moans and screams of Ona.

At last Aniele succeeded in pacifying her, and she essayed the ascent; then, however, she had to be stopped while the old woman cautioned her about the floor of the garret. They had no real floor—they had laid old boards in one part to make a place for the family to live; it was all right and safe there, but the other part of the garret had only the joists of the floor, and the lath and plaster of the ceiling below, and if one stepped on this there would be a catastrophe. As it was half dark up above, perhaps one of the others had best go up first with a candle. Then there were more outcries and threatening, until at last Jurgis had a vision of a pair of elephantine legs disappearing through the trap-door, and felt the house shake as Madame Haupt started to walk. Then suddenly Aniele came to him and took him by the arm.

"Now," she said, "You go away. Do as I tell you—you have done all you can, and you are only in the way. Go away and stay away."

"But where shall I go?" Jurgis asked, helplessly.

"I don't know where," she answered. "Go on the street, if there is no other place—only go! And stay all night!"

Jurgis went outside. It was four o'clock in the morning, and as black as night. There were three or four inches of fresh snow on the ground, and the flakes were falling thick and fast. He turned toward Aniele's and started at a run.

There was a light burning in the kitchen window and the blinds were drawn. The door was unlocked and Jurgis rushed in.

Aniele, Marija, and the rest of the women were huddled about the stove, exactly as before; with them were several newcomers, Jurgis noticed—also he noticed that the house was silent.

"Well?" he said.

No one answered him; they sat staring at him with their pale faces. He cried again: "Well?"

And then, by the light of the smoky lamp, he saw Marija, who sat nearest him, shaking her head slowly. "Not yet," she said.

And Jurgis gave a cry of dismay. "Not yet?"

Again Marija's head shook. The poor fellow stood dumbfounded. "I don't hear her," he gasped.

"She's been quiet a long time," replied the other.

There was another pause—broken suddenly by a voice from the attic: "Hello, there!"

Several of the women ran into the next room, while Marija sprang toward Jurgis. "Wait here!" she cried, and the two stood, pale and trembling, listening. In a few moments it became clear that Madame Haupt was engaged in descending the ladder, scolding and exhorting again, while the ladder creaked in protest. In a moment or two she reached the ground, angry and breathless, and they heard her coming into the room. Jurgis gave one glance at her, and then turned white and reeled. She had her jacket off, like one of the workers on the killing beds. Her hands and arms were smeared with blood, and blood was splashed upon her clothing and her face.

She stood breathing hard, and gazing about her; no one made a sound.

"I haf done my best," she began suddenly. "I can do notting more—dere is no use to try."

Again there was silence.

"It ain't my fault," she said. "You had ought to haf had a doctor, und not vaited so long—it was too late already ven I come." Once more there was a deathlike stillness. Marija was clutching Jurgis with all the power of her one well arm.

Then suddenly Madame Haupt turned to Aniele. "You haf not got someting to drink, hey?" she queried. "Some brandy?"

Aniele shook her head.

"Herr Gott!" exclaimed Madame Haupt. "Such people! Perhaps you vill gif me someting to eat den—I haf not notting since yesterday morning, und I haf vorked myself near to death here. If I could haf known it vas like dis, I vould never haf come for such money as you gif me."

At this moment she chanced to look round, and saw Jurgis. She shook her finger at him. "You understand me," she said, "you pays me dot money yust de same! It is not my fault dat you send for me so late I can't help you vife. It is not my fault if der baby comes mit one arm first, so dot I can't save it. I haf tried all night, und in dot place vere it is not fit for dogs to be born, und mit notting to eat only vot I brings in mine own pockets."

Here Madame Haupt paused for a moment to get her breath; and Marija, seeing the beads of sweat on Jurgis's forehead, and feeling the quivering of his frame, broke out in a low voice: "How is Ona?"

"How is she?" echoed Madame Haupt. "How do you tink she can be ven you leave her to kill herself so? I told dem dot ven they send for de priest. She is young, und she might haf got over it, und been vell and strong, if she been treated right. She fight hard, dot girl—she is not yet quite dead."

And Jurgis gave a frantic scream. "Dead!"

"She vill die, of course," said the other, angrily. "Der baby is dead now."

The garret was lighted by a candle stuck upon a board; it had almost burned itself out, and was sputtering and smoking as Jurgis rushed up the ladder. He could make out dimly in one corner a pallet of rags and old blankets, spread upon the floor; at the foot of it was a crucifix, and near it a priest muttering a prayer. In a far corner crouched Elzbieta, moaning and wailing. Upon the pallet lay Ona.

She was covered with a blanket, but he could see her shoulders and one arm lying bare; she was so shrunken he would scarcely have known her—she was all but a skeleton, and as white as a piece of chalk. Her eyelids were closed, and she lay still as death. He staggered toward her and fell upon his knees with a cry of anguish: "Ona! Ona!"

She did not stir. He caught her hand in his, and began to clasp it frantically, calling: "Look at me! Answer me! It is Jurgis come back—don't you hear me?"

There was the faintest quivering of the eyelids, and he called again in frenzy: "Ona! Ona!"

Then suddenly her eyes opened—one instant. One instant she looked at him—there was a flash of recognition between them, he saw her afar off, as though a dim vista, standing forlorn. He stretched out his arms to her, he called her in wild despair; a fearful yearning surged up in him, hunger for her that was agony, desire that was a new being born within him, tearing his heartstrings, torturing him. But it was all in vain—she faded from him, she slipped back and was gone. And a wail of anguish burst from him, great sobs shook all his frame, and hot tears ran down his cheeks and fell upon her. He clutched her hands, he shook her, he caught her in his arms and pressed her to him; but she lay cold and still—she was gone—she was gone!

QUESTIONS: IDEAS TO THINK ABOUT

1. *What is Sinclair's impression of the midwife?*
2. *How does Sinclair's description of Mrs. Haupt compare with your impressions of Wilhelmine Geipel?*

MRS. GEIPEL'S HORSE AND BUGGY, 1900

Judging by her success, however, the midwife Wilhelmine Geipel of Astoria was a prominent and respected member of the community, quite unlike the unsavory portrait in *The Jungle*. She traveled as far as Sunnyside and Newton, and the horse and carriage she used to go on her rounds served as a symbol of her work and of her status in the community—not unlike a shining new car can serve as a symbol of success and status in our day.

Seated in the horse and buggy in front of the Geipel house are Wilhelmine's sister, Mrs. Emma Hahn, and Dora Nantwich Geipel with their children (Wilhelmine's grandchildren): Dora (with bonnet), August, and Paul. Standing in front of the gate is their maid, Lucy Patersen. In front of the carriage are John Hahn and Minnie Geipel. The barn in the background, converted into a house, is the present home of Dora Geipel, Wilhelmine's granddaughter; the house was torn down in the 1960's.

Wilhelmine Geipel's granddaughter Dora, three years old in the photograph, is now eighty-seven, and in an oral history interview, she talked about her grandmother: "She owned eleven houses in Astoria and one in Corona. She charged ten dollars for each baby she delivered." Dora recollected that she and her sister were often doused with water when they tried to collect the fee for their grandmother's services by people who didn't want to pay Mrs. Geipel for her work.

HORSE AND BUGGY AND THE GEIPEL HOUSE—1900.

QUESTIONS: IDEAS TO THINK ABOUT

1. *How would Mrs. Geipel's life have been changed if she had not had a maid?*
2. *From the information in this photo, how would you describe the Geipels' family life?*

RETURN OF THE MIDWIFE
by Andrea Boroff Eagan

Throughout the twentieth century increasing numbers of women were encouraged by doctors to have their babies in hospitals, and the practice of midwifery was viewed as suitable for only the rural and the poor. Like so many other aspects of modern life, childbirth slipped out of the hands of the "people" and moved under the control of "experts," backed by the authority of science.

Yet midwifery has experienced a Renaissance today, and many women choose to have their babies at home or in special centers where midwives deliver the newborn. The following article from the magazine *Attenzione* (October, 1981) by Andrea Boroff Eagan, tells about a modern woman working as a midwife today.

BY ANDREA BOROFF EAGAN

Back in the old days in Brooklyn's Carroll Gardens, the midwife Angelina D'Amico—with her blue suit, white blouse, and little black bag—was a familiar presence. She was not the only midwife in the neighborhood, but she was popular and delivered many of the babies born there in the first decades of this century. Among her clients was Carmela Fazzia, who had seven babies, all born at home, all delivered by Angelina D'Amico.

Early in each pregnancy, Carmela would start making monthly visits to Angelina for prenatal care. When labor began, Carmela would continue her usual household routine until the pains became frequent and regular enough to summon the midwife. Angelina always came right away—she lived no more than a short walk from any of her clients—and stayed through the long hours of labor until the baby was born. Then the sisters and the husband and the sisters' husbands and the children who had been waiting in the front room would begin the party.

Angelina D'Amico continued to come to the Fazzia household every day for a week, while Carmela stayed in bed. She would wash Carmela and the baby and have a cup of coffee. A new baby was a family event. The women helped out with the housework, the cooking, and in minding the other children; the men partied.

Though most babies were still being born at home as recently as the 1940s, home birth today seems a throwback to a dimly remembered past. What has become routine for us is childbirth as a medical event; drugs, forceps, and even surgery have become commonplace in the management of labor and delivery. By the mid-1960s, however, the natural childbirth movement had begun a trend away from medical intervention, away from drugs in labor. The natural childbirth methods also depended on the woman's having a companion or coach with her throughout labor, and on education and training to help the woman manage her pain. Many doctors and hospitals resisted, afraid that their judgment was being questioned and their control of the childbearing process threatened.

The reasons for the opposition to home birth by most obstetricians, says the Maternity Center's Ruth Lubic, "are complex. It certainly has something to do with wanting to be in control of the birth process. And I would be foolish to rule out the impact of economics, as well as honest-to-God fear. When they get their training in a hospital and see some of the emergencies, all they can say when it's over is thank goodness she was in a hospital. But they don't take account of everything that's happened to that woman."

Unlike many of his colleagues, the physician who works as the backup obstetrician for both the Maternity Center and the Home Birth Center supports the idea of out-of-hospital birth for selected women. The major risks, he believes, are to the newborn, not to the mother, in an out-of-hospital birth. But the things that can happen to newborns, like a drop in the heart rate or a failure to breathe after birth, he notes, "wouldn't generally happen without artificial stimulation of labor, which of course you wouldn't have in the home. They can also happen otherwise, but very, very rarely." Babies born with major genetic defects, of course, must be rushed to a hospital if they are to be treated.

Women who choose home birth are usually aware of the risks of modern hospital practice in obstetrics. Ruth Lubic reports, "The reason many families come to us is not the presence of the technology in the hospital, but its overuse and the untoward things that can happen."

"In the hospital," she continued, "the woman gets her labor adjusted to the hospital, slowed down or speeded up, to accommodate the change of shifts. At home, we have to adapt ourselves and what we do to the woman's specific emotional needs, and her specific family life-style. In the hospital, the woman must adapt to the hospital, and if any members of her family are permitted to be with her, then they too must adapt themselves."

QUESTIONS: IDEAS TO THINK ABOUT

1. *What is the impression of the midwife you get from the Attenzione article?*

2. *As you read the excerpt from the article, consider how the experience of having a baby at home differs from giving birth in a hospital. Suggest pros and cons for each. Why do most women today choose to give birth in hospitals?*

SOJOURNER TRUTH

Wilhelmine Geipel's success in the community was not typical for women at the end of the nineteenth century. Many women argue today that even now, in an age that characterizes itself as liberated, women have enormous problems in the labor force. In fact, women have had difficulties in the work world for centuries. Women have often been expected to do hard manual labor, to raise a family, and to run a household. For a long time they remained second class citizens, deprived of the right to vote and the right to hold property. In 1853, Sojourner Truth, a free black woman, commented on the second class status of women, a situation especially bitter for black women, who are in double jeopardy because of their gender and their race.

SOJOURNER TRUTH.

We Were There by Barbara Wertheimer

Sojourner Truth about 1853. This drawing was the frontispiece of her *Narrative*, which she peddled to support herself as she traveled and preached freedom for blacks and rights for women.

That man over there says that women need to be helped into carriages, and lifted over ditches, and to have the best place everywhere. Nobody ever helps me into carriages, or over mud-puddles, or gives me any best place! And ain't I a woman? Look at me! Look at my arm! I have ploughed, and planted, and gathered into barns, and no man could head me! And ain't I a woman? I could work as much and eat as much as a man—when I could get it—and bear the lash as well! And ain't I a woman? I have borne thirteen children, and seen them most all sold off to slavery, and when I cried out with my mother's grief, none but Jesus heard me! And ain't I a woman?

Then they talk about this thing in the head; what's this they call it? That's it, honey (intellect). What's that got to do with women's rights? . . . If my cup won't hold but a pint and yours hold a quart, wouldn't you be mean not to let me have my little half measure full?

Then that little man in black there, he says women can't have as much rights as men, 'cause Christ wasn't a woman! Where did your Christ come from? *Where did your Christ come from?* From God and a woman! Man had nothing to do with Him.

If the first woman God ever made was strong enough to turn the world upside down all alone, these women together (here) ought to be able to turn it back, and get it right side up again! And now they are asking to do it, the men better let 'em.

QUESTIONS: IDEAS TO THINK ABOUT

1. ***What does Sojourner Truth's statement in the 1860s have in common with the feminist position in the 1980s?***

EXCERPT FROM *A DOLL'S HOUSE*
by Henrik Ibsen, 1868

Sojourner Truth implies the complexities in the relationships between men and women in a world many feel is dominated by men. The sparks of verbal resistance we see in her commentary have smoldered over the centuries, as women have asserted their rights as men's equals. In 1868, just seven years before Wilhelmine Geipel arrived in America and began developing as a key member of the community, Henrik Ibsen, a Swedish playwright, wrote *A Doll's House*. In this popular drama Ibsen shows one woman's escape from a comfortable but stifling domestic setting. The plot traces the story of Nora Torvold's growth from a "little squirrel" in her husband's home, to a forceful, articulate person demanding her right to go out into the world and find meaningful work for herself. We only can wonder about whether or not Wilhelmine and August had to cope with the same issues of roles, duties, and liberation that Ibsen so dynamically presented. But most critics agree that the door that Nora slammed on domesticity still echoes today.

A DOLL'S HOUSE, BY HENRIK IBSEN

HELMER. But what for?—so late as this.

NORA. I shall not sleep tonight.

HELMER. But, my dear Nora—

NORA [looking at her watch]. *It is not so very late. Sit down here, Torvald. You and I have much to say to one another.* [She sits down at one side of the table.]

HELMER. Nora—what is this?—this cold, set face?

NORA. Sit down. It will take some time; I have a lot to talk over with you.

HELMER [sits down at the opposite side of the table]. *You alarm me, Nora!—and I don't understand you.*

NORA. No, that is just it. You don't understand me, and I have never understood you either—before tonight. No, you mustn't interrupt me. You must simply listen to what I say. Torvald, this is a settling of accounts.

HELMER. What do you mean by that?

NORA [after a short silence]. *Isn't there one thing that strikes you strange in our sitting here like this?*

HELMER. What is that?

NORA. We have been married now eight years. Does it not occur to you that this is the first time we two, you and I, husband and wife, have had a serious conversation?

HELMER. What do you mean, serious?

NORA. In all these eight years—longer than that—from the very beginning of our acquaintance we have never exchanged a word on any serious subject.

HELMER. Was it likely that I would be continually and forever telling you about worries that you could not help me to bear?

NORA. I am not speaking about business matters. I say that we have never sat down in earnest together to try and get at the bottom of anything.

HELMER. *But, dearest Nora, would it have been any good to you?*

NORA. *That is just it; you have never understood me. I have been greatly wronged, Torvald—first by Papa and then by you.*

HELMER. *What! By us two—by us two who have loved you better than anyone else in the world?*

NORA [shaking her head]. *You have never loved me. You have only thought it pleasant to be in love with me.*

HELMER. *Nora, what do I hear you saying?*

NORA. *It is perfectly true, Torvald. When I was at home with Papa he told me his opinion about everything, and so I had the same opinions; and if I differed from him I concealed the fact, because he would not have liked it. He called me his doll child, and he played with me just as I used to play with my dolls. And when I came to live with you—*

HELMER. *What sort of an expression is that to use about our marriage?*

NORA [undisturbed]. *I mean that I was simply transferred from Papa's hands to yours. You arranged everything according to your own taste, and so I got the same tastes as you—or else I pretended to. I am really not quite sure which—I think sometimes the one and sometimes the other. When I look back on it it seems to me as if I have been living here like a poor woman—just from hand to mouth. I have existed merely to perform tricks for you, Torvald. But you would have it so. You and Papa have committed a great sin against me. It is your fault that I have made nothing of my life.*

HELMER. *How unreasonable and how ungrateful you are, Nora! Have you not been happy here?*

NORA. *No, I have never been happy. I thought I was, but it has never really been so.*

HELMER. *Not—not happy!*

NORA. *No, only merry. And you have always been so kind to me. But our home has been nothing but a playroom. I have been your doll wife, just as at home I was Papa's doll child; and here the children have been my dolls. I thought it great fun when you played with me, just as they thought it great fun when I played with them. That is what our marriage has been, Torvald.*

HELMER. *There is some truth in what you say—exaggerated and strained as your view of it is. But for the future it shall be different. Playtime shall be over and lesson time shall begin.*

NORA. *Whose lessons? Mine or the children's?*

HELMER. *Both yours and the children's, my darling Nora.*

NORA. *Alas, Torvald, you are not the man to educate me into being a proper wife for you.*

HELMER. *And you can say that!*

NORA. *And I—how am I fitted to bring up the children?*

HELMER. *Nora!*

NORA. *Didn't you say so yourself a little while ago—that you dare not trust me to bring them up?*

HELMER. *In a moment of anger! Why do you pay any heed to that?*

NORA. *Indeed, you were perfectly right. I am not fit for the task. There is another task I must undertake first. I must try and educate myself—you are not the man to help me in that. I must do that for myself. And that is why I am going to leave you now.*

HELMER [springing up]. *What do you say?*

NORA. *I must stand quite alone if I am to understand myself and everything about me. It is for that reason that I cannot remain with you any longer.*

HELMER. *Nora, Nora!*

NORA. *I am going away from here now, at once. I am sure Christine will take me in for the night.*

HELMER. *You are out of your mind! I won't allow it! I forbid you!*

NORA. *It is no use forbidding me anything any longer. I will take with me what belongs to myself. I will take nothing from you, either nor or later.*

HELMER. What sort of madness is this?
NORA. Tomorrow I shall go home—I mean to my old home. It will be easier for me to find something to do there.
HELMER. You blind, foolish woman!
NORA. I must try and get some sense, Torvald.
HELMER. To desert your home, your husband and your children! And you don't consider what people will say!
NORA. I cannot consider that at all. I only know that it is necessary for me.
HELMER. It's shocking. This is how you would neglect your most sacred duties.
NORA. What do you consider my most sacred duties?
HELMER. Do I need to tell you that? Are they not your duties to your husband and your children?
NORA. I have other duties just as sacred.
HELMER. That you have not. What duties could those be?
NORA. Duties to myself.
HELMER. Before all else you are a wife and a mother.
NORA. I don't believe that any longer. I believe that before all else I am a reasonable human being just as you are—or, at all events, that I must try and become one. I know quite well, Torvald, that most people would think you right and that views of that kind are to be found in books; but I can no longer content myself with what most people say or with what is found in books. I must think over things for myself and get to understand them.
HELMER. Can you understand your place in your own home? Have you not a reliable guide in such matters as that?—have you no religion?
NORA. I am afraid, Torvald, I do not exactly know what religion is.
HELMER. What are you saying?
NORA. I know nothing but what the clergyman said when I went to be confirmed. He told us that religion was this and that and the other. When I am away from all this and am alone I will look into that matter too. I will see if what the clergyman said is true, or at all events if it is true for me.
HELMER. This is unheard of in a girl of your age! But if religion cannot lead you aright, let me try and awaken your conscience. I suppose you have some moral sense? Or—answer me—am I to think you have none?
NORA. I assure you, Torvald, that is not an easy question to answer. I really don't know. The thing perplexes me altogether. I only know that you and I look at it in quite a different light. I am learning, too, that the law is quite another thing from what I supposed; but I find it impossible to convince myself that the law is right. According to it a woman has no right to spare her old dying father or to save her husband's life. I can't believe that.
HELMER. You talk like a child. You don't understand the conditions of the world in which you live.
NORA. No, I don't. But now I am going to try. I am going to see if I can make out who is right, the world or I.

HELMER. You are ill, Nora; you are delirious; I almost think you are out of your mind.
NORA. I have never felt my mind so clear and certain as tonight.
HELMER. And is it with a clear and certain mind that you forsake your husband and your children?
NORA. Yes, it is.

QUESTIONS: IDEAS TO THINK ABOUT

1. *In this brief excerpt from* A Doll's House, *Nora explains to her husband, Helmer, what she feels has been wrong with their marriage. What are her main objections?*
2. *Nora and Helmer disagree on the issue of women's "sacred duties." How do you define "sacred duties"? What do you consider women's sacred duties to be?*
3. *Some men today still insist on seeing women as fragile creatures. Why? What jobs are some women still considered too fragile to hold?*
4. *What are the obstacles—financial, social, political—a woman must overcome to work outside the home?*
5. *Are a career and a family mutually exclusive? Is this conflict different for women than for men? Why?*

AUGUST GEIPEL: BREWMASTER TO SALOONKEEPER

We turn from the woman of the Geipel household to the man. When husband and wife arrived in America in 1857, August Geipel too had a marketable skill learned in Germany. Relying, no doubt, upon the active immigrant network of relatives and friends from the Old World who earlier had come to New York— a network is often an immigrant's first source of information about jobs and housing. He quickly found a job at George Ehret's Hell Gate Brewery in Yorkville, just across the East River from Astoria. Geipel's employer, Ehret himself, was an immigrant. He came to America as a young man in 1857; he found his first job in a brewery; in 1866 he went into business for himself. By 1875, when Geipel went to work there, Ehret's was one of the largest and most successful breweries in New York. (Another link between the two families was forged in 1881 when Mrs. Geipel attended at the birth of Ehret's daughter.)

At the brewery August Geipel found an environment with which he was familiar. The work patterns, the customs, the habits, and even the language of the workers were European in origin. Although he worked for Ehret only for about five years, it was long enough for August to establish himself comfortably in the city.

August Geipel, seated on the right, and other brewmasters at the Hell Gate Brewery, c. 1879.

QUESTIONS: IDEAS TO THINK ABOUT

1. *Describe how August might have found his first job. Is the same pattern possible today? What is the relationship between ethnicity and work for him?*
2. *Compare August's situation with his wife's? What were the particular problems each had to face?*

THE GEIPEL SALOON, 1880

For many immigrants an important feature of the "American Dream" was, and is today, owning a neighborhood store: a saloon, a fruit stand, a beauty parlor, or a general store, perhaps catering to the needs of a particular ethnic group.

After working as a foreman in George Ehret's Hell Gate Brewery in Yorkville, August Geipel opened his own saloon in Astoria on Newtown Road in a building owned by his wife. This move to open his own saloon meant that August joined his wife in that both were working for themselves, thus realizing part of the "American Dream."

In his new enterprise he was helped by his friend and former employer, George Ehret; notice the sign for Ehret's lager beer in the saloon window. During the late nineteenth and into the early twentieth century, it was common for breweries, of which there were more than one hundred in New York, to be in the saloon business. Often a brewery offered to provide the would-be saloonkeeper with a building and all of the fixtures for his new business, on the condition, of course, that only their beer was sold.

Success, however, depended on several factors, not the least of which was the personality of the saloonkeeper. A successful saloon like August Geipel's offered in addition to beer and whiskey, food, music, and a place for local clubs, singing societies, and workers' associations to meet.

THE AUGUST GEIPEL WINELAGER BEER SALOON, 1880.

QUESTIONS: IDEAS TO THINK ABOUT

1. *Why is owning a store considered part of the American Dream? Is it as possible today as it was for Geipel? In what ways is it the same? different?*
2. *Was it "ethical" for breweries to operate saloons? Why? or why not? What analogous situation exists today?*
3. *Why do we still have ethnic saloons in New York City? What functions do they still serve? Are there any in your community?*

C.F. HAHN'S STEINWAY AVENUE HOUSE, c. 1880

This photograph shows August Geipel, in the buggy, visiting the saloon of his brother-in-law, Chris Hahn. Emma Hahn, Mrs. Geipel's sister, is in the photo of the horse and buggy. Several details are worth noticing: the horse-car tracks in the street, the signs for Ehret's beer, the group of men holding glasses of foaming dark beer, the women looking out of the second floor window, and the undeveloped land in the distance. The picture also shows how August Geipel's family and professional life intertwined; he is visiting an in-law who is also in the same business. We can only speculate how he and Chris Hahn may have helped each other in his enterprise. Throughout this period Queens remained a county of small shops, family businesses, and farms.

QUESTIONS: IDEAS TO THINK ABOUT

1. *Was the attitude between Geipel and Hahn one of cooperation or competition, or a little of both? Why?*
2. *How do family members help each other in business? Are any of your relatives or in-laws in the same business or profession as you are?*

THE ASSISTANT
by Bernard Malamud

When we consider the idea of free enterprise we must acknowledge the struggle as well as the pride inherent in running one's own business. The store owner's life is one of long hours and hard work; he or she feels a sense of duty towards regular customers, and comes to expect a kind of loyal patronage in return.

Here, from the first chapter of Bernard Malamud's haunting novel *The Assistant,* is a brief portrait of a grocer (which was the most popular occupation in Queens during the late nineteenth century) who endures the rigors of his work and the effects of hard times.

THE ASSISTANT

The early November street was dark though night had ended, but the wind, to the grocer's surprise, already clawed. It flung his apron into his face as he bent for the two milk cases at the curb. Morris Bober dragged the heavy boxes to the door, panting. A large brown bag of hard rolls stood in the doorway along with the sour-faced, gray-haired Poilisheh huddled there, who wanted one.

"What's the matter so late?"

"Ten after six," said the grocer.

"Is cold," she complained.

Turning the key in the lock he let her in. Usually he lugged in the milk and lit the gas radiators, but the Polish woman was impatient. Morris poured the bag of rolls into a wire basket on the counter and found an unseeded one for her. Slicing it in halves, he wrapped it in white store paper. She tucked the roll into her cord market bag and left three pennies on the counter. He rang up the sale on an old noisy cash register, smoothed and put away the bag the rolls had come in, finished pulling in the milk, and stored the bottles at the bottom of the refrigerator. He lit the gas radiator at the front of the store and went into the back to light the one there.

He boiled up coffee in a blackened enamel pot and sipped it, chewing on a roll, not tasting what he was eating. After he had cleaned up he waited; he waited for Nick Fuso, the upstairs tenant, a young mechanic who worked in a garage in the neighborhood. Nick came in every morning around seven for twenty cents' worth of ham and a loaf of bread.

But the front door opened and a girl of ten entered, her face pinched and eyes excited. His heart held no welcome for her.

"My mother says," she said quickly, "can you trust her till tomorrow for a pound of butter, loaf of rye bread and a small bottle of cider vinegar?"

He knew the mother. "No more trust."

The girl burst into tears.

Morris gave her a quarter-pound of butter, the bread and vinegar. He found a penciled spot on the worn counter, near the cash register, and wrote a sum under "Drunk Woman." The total now came to \$2.03, which he never hoped to see. But Ida would nag if she noticed a new figure, so he reduced the amount to \$1.61. His peace—the little he lived with—was worth forty-two cents.

He sat in a chair at the round wooden table in the rear of the store and scanned, with raised brows, yesterday's Jewish paper that he had already thoroughly read. From time to time he looked absently through the square windowless window cut through the wall, to see if anybody had by chance come into the store. Sometimes when he looked up from his newspaper, he was startled to see a customer standing silently at the counter.

Now the store looked like a long dark tunnel.

The grocer sighed and waited. Waiting he thought he did poorly. When times were bad time was bad. It died as he waited, stinking in his nose.

A workman came in for a fifteen-cent can of King Oscar Norwegian sardines.

Morris went back to waiting. In twenty-one years the store had changed little. Twice he had painted all over, once added new shelving. The old-fashioned double windows at the front a carpenter had made into a large single one. Ten years ago the sign hanging outside fell to the ground but he had never replaced it. Once, when business hit a long good spell, he had had the wooden icebox ripped out and a new white refrigerated showcase put in. The showcase stood at the front in line with the old counter and he often leaned against it as he stared out of the window. Otherwise the store was the same. Years ago it was more a delicatessen; now, though he still sold a little delicatessen, it was more a poor grocery.

A half-hour passed. When Nick Fuso failed to appear, Morris got up and stationed himself at the front window, behind a large cardboard display sign the beer people had rigged up in an otherwise empty window. After a while the hall door opened, and Nick came out in a thick, hand-knitted green sweater. He trotted around the corner and soon returned carrying a bag of groceries. Morris was now visible at the window. Nick saw the look on his face but didn't look long. He ran into the house, trying to make it seem it was the wind that was chasing him. The door slammed behind him, a loud door.

The grocer gazed into the street. He wished fleetingly that he could once more be out in the open, as when he was a boy—never in the house, but the sound of the blustery wind frightened him. He thought again of selling the store but who would buy? Ida still hoped to sell. Every day she hoped. The thought caused him grimly to smile, although he did not feel like smiling. It was an impossible idea so he tried to put it out of his mind. Still, there were times when he went into the back, poured himself a spout of coffee and pleasantly thought of selling. Yet if he miraculously did, where would he go, where? He had a moment of uneasiness as he pictured himself without a roof over his head. There he stood in all kinds of weather, drenched in rain, and the snow froze on his head. No, not for an age had he lived a whole day in the open. As a boy, always running in the muddy, rutted streets of the village, or across the fields, or bathing with the other boys in the river; but as a man, in America, he rarely saw the sky. In the early days when he drove a horse and wagon, yes, but not since his first store. In a store you were entombed.

The milkman drove up to the door in his truck and hurried in, a bull, for his empties. He lugged out a caseful and returned with two half-pints of light cream. Then Otto Vogel, the meat provisions man, entered, a bushy-mustached German carrying a smoked liverwurst and string of wieners in his oily meat basket. Morris paid cash for the liverwurst; from a German he wanted no favors. Otto left with the wieners. The bread driver, new on the route, exchanged three fresh loaves for three stale and walked out without a word. Leo, the cake man, glanced hastily at the package cake on top of the refrigerator and called, "See you Monday, Morris."

Morris didn't answer.

Leo hesitated. "Bad all over, Morris."

"Here is the worst."

"See you Monday."

A young housewife from close by bought sixty-three cents' worth; another came in for forty-one cents'. He had earned his first cash dollar for the day.

Bernard Malamud

QUESTIONS: IDEAS TO THINK ABOUT

1. *What are some of the problems Morris has trying to run his store?*
2. *What does Morris feel the store did to his young manhood?*
3. *In what kind of neighborhood is this store located? What tells you that in this passage?*
4. *What is this shopkeeper's role in his community?*
5. *How do small shop owners survive in New York City today? What kinds of problems do they face?*

THE BONNER-CANNON-AARON FAMILY

At about the same time as the Geipel family was settling in New York as immigrants from Germany, James Bonner, a black sharecropper in Virginia, decided to move north with his wife, Nancy, and their daughter Eleanor. They were one of many black families who left the South in the 1870's and 1880's for the opportunity and promise of life in northern cities.

We will trace the work history of this family over three generations, from the Bonners' arrival in Queens to the story of their granddaughter, Nancy Aaron. An important theme which runs through their story is how the family members relied on one another for work and moral support. As you examine the documents and information on the following pages, it will be helpful to keep in mind this family work-tree.

James and Nancy Bonner
(Civil War Veteran, Sharecropper, Shopkeeper)

Eleanor Bonner *(Waitress, Dressmaker)* m. Smith Cannon *(Bricklayer, Cook)*

Nancy Cannon *(Domestic, Factory worker)* m. George Aaron *(Meat packer, Soldier, Handyman)*

THE BLACK EXPERIENCE UNDER SLAVERY

To understand the reasons why James Bonner wanted to come north with his family, it is necessary to look back in time. Bonner fought for the Union Army during the Civil War, and even if he was not born into slavery it is likely that his parents were. Therefore, the experience of Afro-Americans under slavery is central to our understanding of black history and culture.

Of all the immigrant groups that came to America to live and work, only blacks came as slaves. As a labor system, slavery was less profitable, more difficult to sustain, and in many ways more brutal than "free" labor in the north. Perhaps the most important aspect of the history of slavery in America is how blacks survived under conditions of brutality and repression, their forms of resistance, and the attitudes they developed in response to their harsh treatment. Slaves did not merely imitate the ways of their white masters, but developed original patterns of family and community life.

Without a written history of their own, slaves developed a lively oral tradition to transmit their story from one generation to the next. *Aunt Sue's Stories*, by the black poet Langston Hughes, is in a sense a poem about oral history. In both slavery and freedom, every family had their "Aunt Sue," the oldest member of the extended family or the community who was respected and looked to for advice and authority. She was also the keeper of their past.

AUNT SUE'S STORIES
BY LANGSTON HUGHES

Aunt Sue has a head full of stories.
Aunt Sue has a whole heart full of stories.
Summer nights on the front porch
Aunt Sue cuddles a brown-faced child to her bosom
And tells him stories.

Black slaves
Working in the hot sun,
And black slaves
Walking in the dewy night,
And black slaves
Singing sorrow songs on the banks of a mighty river
Mingle themselves softly
In the flow of old Aunt Sue's voice,
Mingle themselves softly
In the dark shadows that cross and recross
Aunt Sue's stories.

And the dark-faced child, listening,
Knows that Aunt Sue's stories are real stories.
He knows that Aunt Sue never got her stories
Out of any book at all,
But that they came
Right out of her own life.

The dark-faced child is quiet
Of a summer night
Listening to Aunt Sue's stories.

QUESTIONS: IDEAS TO THINK ABOUT

1. *Who in your family is an "Aunt Sue"?*
2. *Who first told you stories of family history?*
3. *What are some of your family's stories?*

RECONSTRUCTION IN THE SOUTH

People like James and Nancy Bonner came north with their families in the 1870's for a number of reasons. After the Civil War, during the period known as Reconstruction, the south was administered by northerners and occupied by the Union Army. Attempts were made to improve the legal situation of the former slaves, culminating in the Civil Rights Act of 1875 (decalred unconstitutional in 1883) which guaranteed to blacks "the full and equal enjoyment of all public facilities." Even so, the south was rife with lynchings and other attacks on their freedom. Through economic and political pressure on the freed-men, and the institution of "Jim Crow" laws, a separate and unequal society was recreated. With the rise of the Ku Klux Klan in the 1890's the situation for blacks worsened.

The paintings reproduced here, though depicting the south in the twentieth century, give vivid impressions of the conditions blacks faced, the economic despair on the one hand, and the terror of lynchings on the other.

LYNCHING, 1939. Arnold Blanch (1896-1968).

GEORGIA JUNGLE, 1939. Alexander Brook (1898-1980).

QUESTIONS: IDEAS TO THINK ABOUT

1. *From the two paintings, what is your impression of life in the South for blacks? Do you think it is accurate? Why or why not?*

MORTGAGE, 1886

It is unusual and significant to find a record of this kind, for it contradicts the impression that the blacks who came north were poor and unemployed, and the implication that they could not better themselves. Here, within a short time after he migrated, Bonner was able to buy his own house. There is not yet enough information to conclude whether this was unusual or not, but it does force us to question our former assumptions.

In 1886, James Bonner bought a 25' x 100' plot of land on Marion Street (now 13th Street) in Long Island City. He paid $175 for the property, and he signed an Indenture agreement stating that if he were unable to make the payments of $10 "to be paid on the 2nd day of each and every month," he would have to pay a "penal sum" of twice what he owed.

There was also the interest of "six percent per annum, to be paid semiannually." In fact, the mortgage contract contained several sections that were really hidden costs: interest, tax, assessment, and insurance clauses. In James Bonner's mortgage, one clause called for him to insure his property. If he failed to do so, the seller was allowed by law to insure it for him. The cost of the premium, plus 6% interest, was simply added to the monthly payment.

The penalties spelled out in the contract also show the hard terms that the seller was able to demand, and no doubt these conditions soured Bonner's pleasure somewhat. Historically, these agreements mark the beginning of finance companies and installment buying for consumers. Then as now, the companies were not above setting stiff conditions for the sale, or taking advantage of people who did not understand what they were signing.

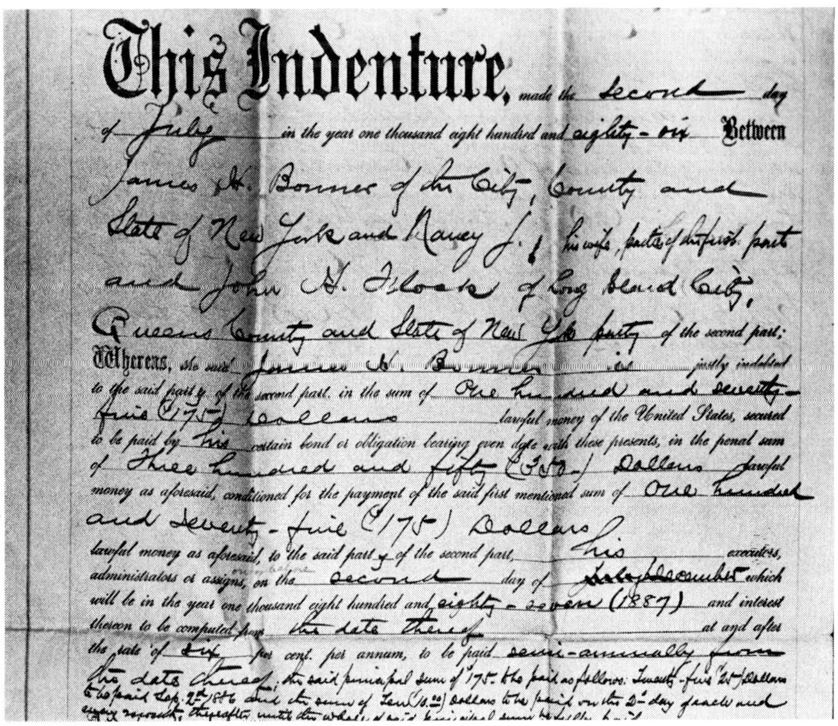

This Indenture, made the second day of July in the year one thousand eight hundred and eighty-six Between James H. Bonner of the City, County and State of New York and Nancy J., his wife, parties of the first part and John H. Hloak of Long Island City, Queens County and State of New York party of the second part; Whereas, the said James H. Bonner is justly indebted to the said party of the second part, in the sum of One hundred and seventy-five ($175) Dollars lawful money of the United States, secured to be paid by his certain bond or obligation bearing even date with these presents, in the penal sum of Three hundred and fifty ($350) Dollars lawful money as aforesaid, conditioned for the payment of the said first mentioned sum of One hundred and seventy-five ($175) Dollars lawful money as aforesaid, to the said party of the second part, his executors, administrators or assigns, on the second day of December which will be in the year one thousand eight hundred and eighty-seven (1887) and interest thereon to be computed from the date thereof at and after the rate of six per cent. per annum, to be paid semi-annually from the date thereof, the said principal sum of 175 to be paid as follows: Twenty-five ($25) dollars to be paid Sep. 2nd 1886 and the sum of Ten ($10.00) dollars to be paid on the 2nd day of each and every month thereafter until the whole of said principal sum [illegible]

BONNER MORTGAGE—1886. "Cheap as the houses were, they were sold with the idea that the people who bought them would not be able to pay for them. When they failed—if it were only by a single month—they would lose the house and all that they had paid on it, and then the company would sell it over again." — Upton Sinclair, The Jungle.

QUESTIONS: IDEAS TO THINK ABOUT

1. *How might the seller have taken advantage of James Bonner?*
2. *Is the buyer protected today? What are some unscrupulous business practices finance companies engage in today?*
3. *Discuss your impressions of a contract you signed recently.*

BIRTH CERTIFICATE

When Mr. Bonner first came north, he opened his own restaurant in Manhattan, and his daughter Eleanor often helped out. In Nancy Aaron's words: "He had a little restaurant somewheres downtown, in the 50's or 60's. My father went there to eat, and that's how he met my mother." Eleanor married Smith Washington Cannon shortly after they met, and they moved next door to her parents at 72 Marion Street in Long Island City, where their first child, Nancy, was born.

Three points stand out on the birth certificate reproduced here. First, the mother had her child at home, possibly assisted by a midwife. Second, the word "colored" was the official designation of blacks, and the category is "color" rather than "race." Third, the father's occupation is listed, while the mother's is left out.

Smith Cannon had been working as a bricklayer, but after marriage he went to work as a cook in his father-in-law's restaurant. His wife became a dressmaker, working out of their home. Again, from Nancy Aaron: "My mother used to work for rich people. There were quite a few rich people in those days, quite a few poor people, too. And the poor people worked for the rich people. ... She used to make dresses for anybody who wanted a dress made. She was a good dressmaker and everyone recommended her to someone else. Just kept busy making clothes, making dresses."

It is important to note the combination of workplace and community, and how Smith Cannon was immediately embraced as a member of both. With the Bonners and Cannons living next to each other and working together, there was an increased chance of them all prospering. Later, the pattern is repeated when Nancy's husband is brought into the fold. He worked out of the home much as Nancy's mother did, obtaining work by word of mouth through friends and neighbors.

QUESTIONS: IDEAS TO THINK ABOUT

1. *What does this document tell you about the Cannons?*
2. *Look at your own birth certificate. How is it different from Nancy Cannon's birth certificate?*

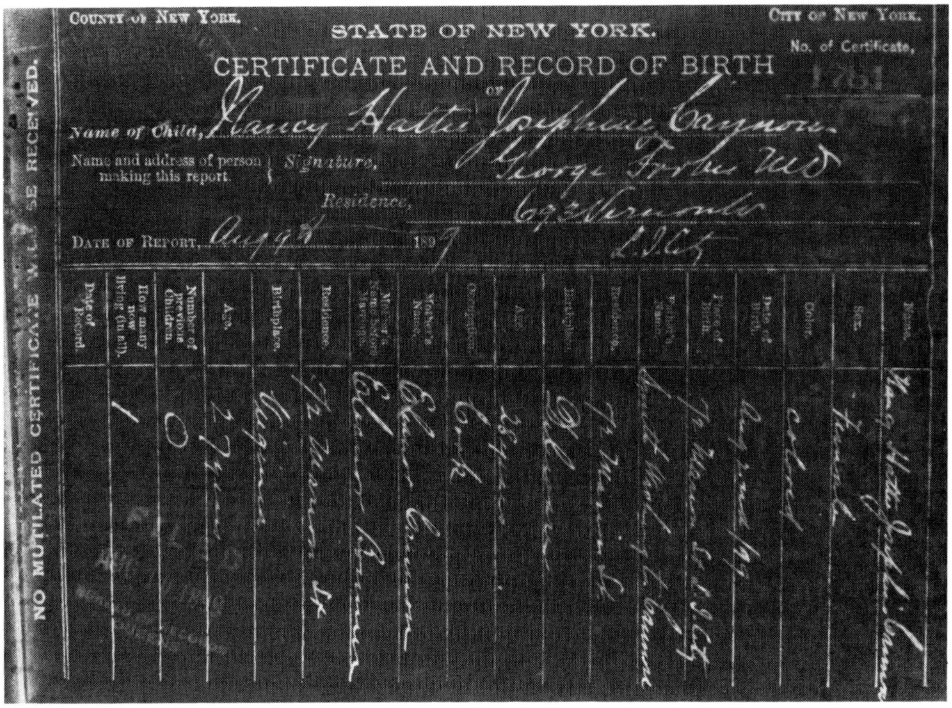
COUNTY OF NEW YORK.
STATE OF NEW YORK.
CITY OF NEW YORK.
No. of Certificate,
CERTIFICATE AND RECORD OF BIRTH
of
Name of Child,
Name and address of person making this report. Signature,
Residence,
DATE OF REPORT, 189
NO MUTILATED CERTIFICATE WILL BE RECEIVED.

BIRTH CERTIFICATE: NANCY H.J. CANNON (1899).

SHOE FACTORY

Although Nancy Cannon's first job was as a domestic—she went to work with her cousin, a cook, in New Jersey during her summer vacation—she returned to Queens and took a job at the Comfort Sandal Manufacturing Company. She had intended to return to school.

Nancy was 15 years old when she went to work at the shoe factory, which was within walking distance of her home, and she stayed there for 14 years. Only three other Black women worked there at that time. Most of the workers were Italian or German. Nancy recalls no resentment or segregation, however: "I didn't know I was black until I looked at myself."

For her first week's work she received $5 for a six day week. When she quit, she was getting $30-$35 for a five day, 40 hour week. She was paid in cash, once a week: "... come home, go to grocer's, go to butcher's, that's all you could do."

In 1919, Nancy Cannon married George Aaron. She stopped working at the shoe factory when her son, George, was born in 1928. When he was old enough, she went to work at the Queens Laundry on Webster Avenue. The laundry, like the factory, was within walking distance of her home. "I didn't get too much, about $20 or $25 a week," she says. "It's just the idea of working."

Nancy Aaron has lived and worked in Long Island City all her life. Today she still lives in a house on the original plot of land that her grandfather, James Bonner, bought in 1886.

Comfort Sandal Manufacturing Co.

<u>Comfort Sandal Manufacturing Company</u>

This concern was established in 1912 and moved to Queens in 1916 and located at 87-93 Sunswick Street, Long Island City. The building is of steel and concrete, five stories and basement and contains 27,000 square feet of space.

They employ 110 persons and the daily capacity of the plant is approximately 3,000 pairs of stitch-downs, shoes, oxfords and sandals. They specialize in the "comfort-wear" brand of shoes. The firm is composed of J. Fischel and M. Simon.

-Queensborough Chamber of Commerce Yearbook, 1927.

SHOE FACTORIES IN QUEENS BOROUGH

Name	Year Established in Queens	Space Occupied Square Feet	Number of Employees
Frank Barber Shoe Co.	1914	46,000	300
Bally, Inc.	1915	27,000	135
Comfort Sandal Mfg. Co.	1916	27,000	110
A. Garside & Sons, Inc.	1917	85,000	250
Frank Pesl	1918	20,000	100
A. B. Means & Co.	1921	53,000	150
Five Star Shoe Co., Inc.	1921	10,000	75
Rubin Brothers	1923	70,000	500
P. & B. Rubber Company	1923	5,000	20
I. Miller & Sons	1924	150,000	1,400
Philipson-Lockwood, Inc.	1927	35,000	200
Crown Legging Co.	1927	6,000	50
Rosenwasser Shoe Corp.	1927	13,500	65
		547,500	3,355

"The air in the ... establishment strangled her. She knew she was gradually and surely shrivelling in the hot, stuffy room."

-Stephen Crane, <u>Maggie: A Girl of the Streets.</u>

". . . instead of going to school I went to the shoe factory cause my sister worked in the shoe factory and she was making the big money, $5.00 a week. . . I used to lace the shoes, polish the shoes, cut the little strings off."

-Nancy Aaron, Oral History.

WOMEN IN FACTORIES: TWO POEMS

These poems by Lucy Larcom and Carl Sandburg give a sense of the despair felt by women, girls really, who took jobs in factories when they were quite young, thus surrendering their youth for the life —more like a slow death—of a factory worker. In Carl Sandburg's words: "You are old before you are young." Lucy Larcom speaks of the effects of "binding shoes" on a young girl, a "bright-eyed beauty once," but now "faded, wrinkled."

THE LOWELL OFFERING

by Lucy Larcom, c. 1843

Poor lone Hannah
Sitting at the window binding shoes;
Faded, wrinkled,
Sitting stitching in a mournful muse.
Bright-eyed beauty once was she,
When the bloom was on the tree;
Spring and winter,
Night and morning,
Hannah's at the window, binding shoes.

MILL DOORS

by Carl Sandburg

You never come back.
I say good-bye when I see you going in the doors,
The hopeless open doors that call and wait
And take you then for-how many cents a day?
How many cents for the sleepy eyes and fingers?

I say good-bye because I know they tap your wrists,
In the dark, in the silence, day by day,
And all the blood of you drop by drop,
And you are old before you are young.
You never come back.

QUESTIONS: IDEAS TO THINK ABOUT

1. *What do these poems tell you about working conditions in factories?*
2. *"The Lowell Offering" was written and printed by women who worked in the mills. What do you imagine those women were like? Do workers produce, anything like the "Lowell Offering" today?*
3. *How are conditions in factories today different from or the same as those described in the poems? Have the attitudes of factory workers toward their tasks changed?*

WOMEN IN THE STEAM LAUNDRIES

This selection from Barbara Wertheimer's *We Were There: The Story of Working Women in America* gives us further insight into factory conditions, especially for women working in unbearably hot and unsanitary laundries. The wages paid these women were much lower than those paid to men.

THE IRONERS, 1943. Jacob Lawrence (b. 1917).

WOMEN IN THE STEAM LAUNDRIES

How would you like to iron a shirt a minute? Think of standing at a mangle just above the washroom with the hot steam pouring up through the floor for 10, 12, 14 and sometimes 17 hours a day! Sometimes the floors are made of cement and then it seems as though one were standing on hot coals, and the workers are dripping with perspiration. Perhaps you have complained about the chemicals used in the washing of your clothes, which cause them to wear out quickly, but what do you suppose is the effect of these chemicals upon the workers? They are . . . breathing air laden with particles of soda, ammonia, and other chemicals! The Laundry Workers Union . . . in one city reduced this long day to 9 hours, and has increased the wages 50 percent. . . .

Investigations turned up not only low wages and workweeks 78 hours long, but found that women were fired for refusing to perform overtime at no pay. Illinois factory inspector Florence Kelley reported women fainting at their work: "Girls have been removed from the laundry to the hospital suffering exhaustion after working sixteen, eighteen, and even twenty hours in heat and dampness in ill-ventilated laundries."

In San Francisco in 1900, laundry workers "lived in," each laundry providing board and lodging for its employees.

> The dormitories were stretched places, with four beds in each small room. The food was poor and scanty, and even though the girls worked until midnight or after, no food was allowed after the evening meal at 6 o'clock. Half an hour only was allowed for lunch. Early in the morning, the women were routed out in no gentle manner and by 6 o'clock the unwholesome breakfast was over, and every one hard at work. . . .

These laundry workers earned no more than $8.10 in cash wages each month, in addition to their room and board. Those who lived at home could expect from $17.50 to $25.00 a month. So many letters of complaint were received by newspapers and by the California State Labor Commissioner about conditions in these laundries that the state decided to investigate. Inspectors found violations of the law forbidding work after 10:00 P.M., with women often on the job until 2:00 in the morning. As a result a new ordinance was passed forbidding work after 7:00 at night.

By this time laundry workers had realized that no law would be enforced unless they themselves got together to demand it. Three hundred women applied to the Laundry Workers International Union for a charter. Although the men in the laundries did not want women in their union, to its credit the parent International Union insisted that the women be admitted.

The organizing campaign was conducted secretly, as Hannah Mahony (later Nolan) realized that many of the women were too intimidated by the employers to join the new union openly. Within sixteen weeks a majority had signed up, along with the men. Not until the union was strong did it announce its demands: elimination of the boardinghouse system, shorter hours, wages of $6 to $10 a week, nine holidays a year, time-and-a-half for overtime, and a regular lunch hour with a fine imposed on those not taking it.

By the time these demands were presented to the employers, it was too late for them to fight the union. Among later gains that workers won was the 9-hour day. For years this laundry union thrived, even surviving the tragic San Francisco earthquake and fire of 1906, which destroyed its union headquarters and two thirds of the city's laundries.

In 1907 the union presented new demands for the 8-hour day and higher wages. When the owners refused, 1,100 workers from fourteen laundries walked out. The union was able to hold out for eleven weeks, supporting its members, and reached a compromise settlement that brought workers the 51-hour week, to be reduced gradually until the 8-hour day was in place by April of 1910. There was to be no work before seven in the morning. All work after 5:00 at night was to receive overtime, except on Mondays, when laundries worked until 6:00. By 1912 the San Francisco laundry workers had brought up wages by 30 percent, installed safety measures in the plants, and organized all the city's steam laundries. Few unions in such a low-wage, unskilled industry had ever lasted long or accomplished so much.

Barbara Wertheimer

THE WOMAN WARRIOR

by Maxine Hong Kingston

Conditions in laundries and other sweat shops still leave much room for improvement. The selection below from *The Woman Warrior*, written by a Chinese-American, Maxine Hong Kingston, recalls the grueling hours and hard work her mother endured in the family laundry. The mother here, like Nancy Aaron, has to balance her work in the laundry and her job as child raiser. The writer also compares work conditions in China and work conditions in America. (Note: When Kingston refers to "ghosts," she is referring to her white customers.)

THE WOMAN WARRIOR

"I didn't need muscles in China. I was small in China." She was. The silk dresses she gave me are tiny. You would not think the same person wore them. This mother can carry a hundred pounds of Texas rice up- and downstairs. She could work at the laundry from 6:30 a.m. until midnight, shifting a baby from an ironing table to a shelf between packages, to the display window, where the ghosts tapped on the glass. "I put you babies in the clean places at the laundry, as far away from the germs that fumed out of the ghosts' clothes as I could. Aa, their socks and handkerchiefs choked me. I cough now because of those seventeen years of breathing dust. Tubercular handkerchiefs. Lepers' socks." I thought she had wanted to show off my baby sister in the display window.

In the midnight unsteadiness we were back at the laundry, and my mother was sitting on an orange crate sorting dirty clothes into mountains—a sheet mountain, a white shirt mountain, a dark shirt mountain, a work-pants mountain, a long underwear mountain, a short underwear mountain, a little hill of socks pinned together in pairs, a little hill of handkerchiefs pinned to tags. Surrounding her were candles she burned in daylight, clean yellow diamonds, footlights that ringed her, mysterious masked mother, nose and mouth veiled with a cowboy handkerchief. Before undoing the bundles, my mother would light a tall new candle, which was a luxury, and the pie pans full of old wax and wicks that sometimes sputtered blue, a noise I thought was the germs getting seared.

"No tickee, no washee, mama-san?" a ghost would say, so embarrassing.

"Noisy Red-Mouth Ghost," she'd write on its package, naming it, marking its clothes with its name.

Back in the bedroom I said, "The candles must have helped. It was a good idea of yours to use candles."

"They didn't do much good. All I have to do is think about dust sifting out of clothes or peat dirt blowing across a field or chick mash falling from a scoop, and I start coughing." She coughed deeply. "See what I mean? I have worked too much. Human beings don't work like this in China. Time goes slower there. Here we have to hurry, feed the hungry children before we're too old to work. I feel like a mother cat hunting for its kittens. She has to find them fast because in a few hours she will forget how to count or that she had any kittens at all. I can't sleep in this

country because it doesn't shut down for the night. Factories, canneries, restaurants—always somebody somewhere working through the night. It never gets done all at once here. Time was different in China. One year lasted as long as my total time here; one evening so long, you could visit your women friends, drink tea, and play cards at each house, and it would still be twilight. It even got boring, nothing to do but fan ourselves. Here midnight comes and the floor's not swept, the ironing's not ready, the money's not made. I would still be young if we lived in China."

"Time is the same from place to place," I said unfeelingly. "There is only the eternal present, and biology. The reason you feel time pushing is that you had six children after you were forty-five and you worried about raising us. You shouldn't worry anymore, though, Mama. You should feel good you had so many babies around you in middle age. Not many mothers have that. Wasn't it like prolonging youth? Now wasn't it? You mustn't worry now. All of us have grown up. And you can stop working."

"I can't stop working. When I stop working, I hurt. My head, my back, my legs hurt. I get dizzy. I can't stop."

"I'm like that too, Mama. I work all the time. Don't worry about me starving. I won't starve. I know how to work. I work all the time. I know how to kill food, how to skin and pluck it. I know how to keep warm by sweeping and mopping. I know how to work when things get bad."

"It's a good thing I taught you children to look after yourselves. We're not going back to China for sure now."

"You've been saying that since nineteen forty-nine."

"Now it's final. We got a letter from the villagers yesterday. They asked if it was all right with us that they took over the land. The last uncles have been killed so your father is the only person left to say it is all right, you see. He has written saying they can have it. So. We have no more China to go home to."

Maxine Hong Kingston

QUESTIONS: IDEAS TO THINK ABOUT

1. *From these two selections, what are your impressions of work in steam laundries?*
2. *What is the role of ethnicity in The Woman Warrior?*
3. *In what ways does the mother feel work conditions are different in China? Why do you think the mother ends this selection by stating, "We have no more China to go home to"?*
4. *Why does Kingston refer to white people as "ghosts"?*

BLACK MEN IN THE ARMED FORCES

George Aaron, Nancy's future husband, was born in 1896 in South Carolina, and he served in the First World War. Although there is no indication of "color" on his Enlistment Record, extensive segregation did exist in the armed forces until after World War II.

During World War I, as members of the armed forces, blacks were allowed to serve in the Navy only as messboys and were barred entirely from the Marines. The Army accepted enlisted men, but planned originally not to commission blacks as officers. Only after action by black college students, NAACP officials, and a committee of prominent white citizens did the War Department finally establish a negro officers' training camp at Des Moines, Iowa. In October, 1917, 639 men were commissioned with the ranks of captain and first and second lieutenants; later, others became officers of unsegregated officer-training camps and in the field. All served only as officers of all-black units, however.

During the Second World War, conditions for blacks were much better than they had been in the First World War. As in the First World War, however, major problems continued: employment, racial tensions in the cities and at army camps, and the mistreatment of black soldiers. Although blacks were sent to integrated army officers' training camps, in the lower ranks of the armed forces they were segregated. Conditions changed radically with the admission of negroes into the Army Corps and the Marines, and for the first time since early in the century they were accepted in naval grades above messboy. On the naval posts, though, recreational facilities were segregated, and there were reports from Italy of low morale because of the attitudes of prejudiced white army officers. In the south, black servicemen were often insulted by white citizens, a situation dramatized in Frank Yerby's short story on the following pages.

George Aaron's enlistment record shows what information the army thought was important and it gives clues as to the position of blacks in the armed forces. Before he enlisted, George Aaron had been a meat packer, and in the army he served in a service company. He was "not qualified" to be a gunner and did not become a noncommissioned officer. He served with the American Expeditionary Force in France from September 20, 1918, less than two months from the day he enlisted, until June, 1919. It should be noted that during World War I black regiments were usually the first ones sent into battle and suffered the highest casualties. They also remained in France long after most of the other American soldiers had been sent home.

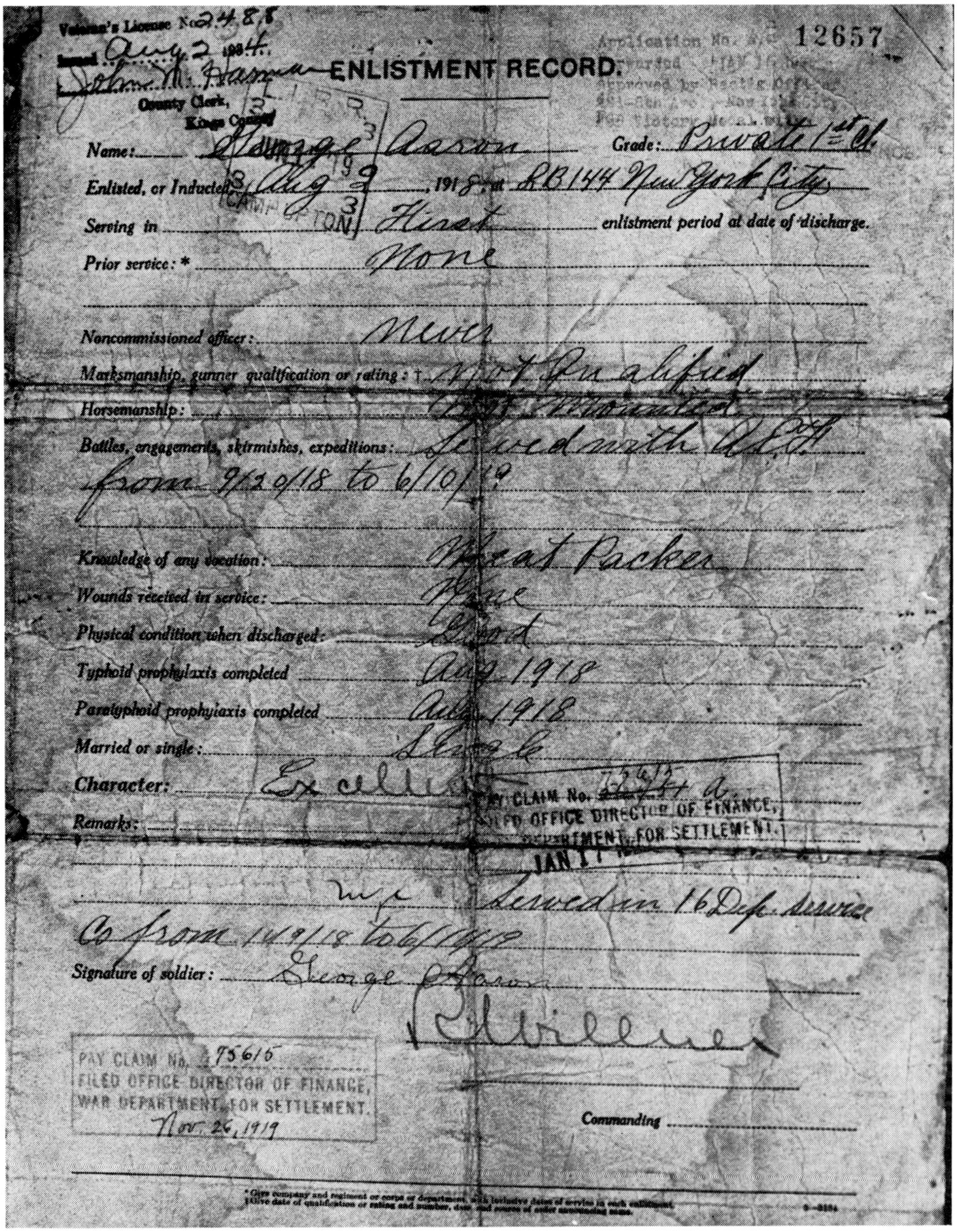

Veteran's License No. 2488
Issued Aug 2 1934
John M. Hasman
County Clerk,
Kings County

Application No. 12657

ENLISTMENT RECORD.

Name: George Aaron Grade: Private 1st Cl.

Enlisted, or Inducted Aug 9, 1918, at LB144 New York City

Serving in First enlistment period at date of discharge.

Prior service: * None

Noncommissioned officer: Never

Marksmanship, gunner qualification or rating: † Not Qualified

Horsemanship: Not Mounted

Battles, engagements, skirmishes, expeditions: Served with A.E.F. from 9/20/18 to 6/10/19

Knowledge of any vocation: Meat Packer

Wounds received in service: None

Physical condition when discharged: Good

Typhoid prophylaxis completed Aug 1918

Paratyphoid prophylaxis completed Aug 1918

Married or single: Single

Character: Excellent

PAY CLAIM No.
FILED OFFICE DIRECTOR OF FINANCE,
WAR DEPARTMENT, FOR SETTLEMENT.
JAN 17

Remarks: Served in 16 Dep. Service Co from 10/9/18 to 6/1919

Signature of soldier: George Aaron

PAY CLAIM No. 75616
FILED OFFICE DIRECTOR OF FINANCE,
WAR DEPARTMENT, FOR SETTLEMENT.
Nov. 26, 1919

Commanding

ENLISTMENT RECORD: GEORGE AARON.

QUESTIONS: IDEAS TO THINK ABOUT

1. *What does this record suggest about the treatment of Blacks in the army during World War I?*
2. *How might the wartime experience have affected Aaron and other Black soldiers?*
3. *What happens to a community when the men go off to war? What problems do the men face when they return?*

SONNET TO NEGRO SOLDIERS
by Joseph Seaman Cotter, Jr.

They shall go down unto Life's Borderland,
Walk unafraid within that Living Hell
Nor heed the driving rain of shot and shell
That round them falls; but with uplifted hand
Be one with mighty hosts, an armed band
Against man's wrong to man—for such full well
They know. And from their trembling lips shall swell
A song of hope the world can understand.

All this to them shall be a glorious sign,
A glimmer of that resurrection morn
When age-long faith, crowned with a grace benign,
Shall rise and from their brows cast down the thorn
Of prejudice. E'en though through blood it be,
There breaks this day their dawn of liberty.

THE HOMECOMING
by Frank Yerby

What did the Black soldier find after returning from the war, after fighting for freedom overseas? Unfortunately, he encountered a society holding the same oppressive attitudes and offering the same limited opportunities as when he left to fight. The soldier had proven himself in battle, but that meant little in civilian life.

In his *Sonnet to Negro Soldiers*, Joseph Seaman Cotter, Jr. (1895-1919) expressed the feeling of the Black soldier that by fighting "Against man's wrong to man" in the first World War he was triumphing over the racial prejudice he had endured at home.

In a more realistic, but depressing vein, Frank Yerby's short story, "The Homecoming," describes what a black soldier could expect on his return home in the South. The story is about a soldier who returns from the war with an artificial leg, and his realization that in the eyes of the "white folks" he is a negro first and a brave veteran last.

"THE HOMECOMING"

He started across the street toward the Monument, throwing one leg a little stiffly. The white men who always sat around it on the little iron benches looked at him curiously. He came on until he stood in the shadow of the shaft. He looked up at the statue of the Confederate soldier, complete with knapsack and holding the musket with the little needle-type bayonet ready for the charge. At the foot of the shaft there was an inscription carved in stone. Willie spelled out the words:

"No nation rose so white and pure; none fell so free of stain."

He stood there, letting the words sink into his brain.

One of the tall loungers took a sliver of wood out of his mouth and grinned. He nudged his companion.

"What do it say, boy?" he asked.

Willie looked past him at the dusty, unpaved streets straggling out from the Monument.

"I ask you a question, boy." The white man's voice was very quiet.

"You talking to me?" Willie said softly.

"You know Goddamn well I'm talking to you. You got ears, ain't you?"

"You said boy," Willie said. "I didn't know you was talking to me."

"Who the hell else could I been talking to, nigger?" the white man demanded.

"I don't know," Willie said. "I didn't see no boys around."

The two white men got up.

"Aint' you forgetting something, nigger?" one of them asked, walking toward Willie.

"Not that I knows of," Willie declared.

"Ain't nobody ever told you to say sir to a white man?"

"Yes," Willie said. "They told me that."

"Yes what?" the white man prompted.

"Yes nothing." Willie said quietly. "Jus plain yes. And I don't think you better come any closer, white man."

"Nigger, do you know where you're at?"

"Yes," Willie said. "Yes, I knows. And I knows you can have me killed. But I don't care about that. Long time now I don't care. So please don't come no closer, white man. I'm asking you kindly."

The two men hesitated. Willie started toward them, walking very slowly. They stood very still, watching him come. Then at the last moment, they stood aside and let him pass. He limped across the street and went into the town's lone Five and Ten Cent Store.

Later, Willie visits his former white benefactor, Colonel Bob, who can't understand how the war has changed him. When Willie says, "Got to be a man. My own man. . . . Can't ask for handouts round the back door. Got to come in the front door . . . ," Colonel Bob silently motions for him to leave. At that moment word comes that there is trouble in town about "some colored soldier."

Willie stood very still, looking up through a break in the trees at the pale, whitish blue sky. It was very high and empty. And in the trees, no bird sang. But Colonel Bob was coming back now, his face very red, and knotted into hard lines.

"Willie," he said, "did you tell two white men you'd kill them if they came nigh you?"

"Yes. I didn't say that, but that's what I meant."

"And did you have some kind of an argument with a white woman?"

"Yes, Colonel Bob."

"My God!"

"He's crazy, Colonel Bob," Martha wailed. "He done gone plum outa his mind!"

"You better not go back to town," the Colonel said. "You better stay here until I can get you out after dark."

Willie smiled a little.

"I'm gonna ketch me a train," he said. "Two o'clock today, I'm gonna ketch it."

"You be kilt!" Martha declared. "They kill you sure!"

"We done run too much, Martha," Willie said slowly. "We done run and hid and anyhow we done got caught. And then we goes down on our knees and begs. I ain't running. Done forgot how. Don't know how to run. Don't know how to beg. Just knows how to fight, that's all, Martha."

"Oh, Jesus, he crazy! Told you he crazy, Colonel Bob!"

Colonel Bob was looking at Willie, a slow, thoughtful look.

"Can't sneak off in the dark, Colonel Bob. Can't steal away to Jesus. Got to go marching. And don't a man better touch me." He turned and went down the steps. "Good-by Colonel Bob," he called.

"Crazy," Martha wept. "Out of his mind!"

"Stop your blubbering!" Colonel Bob snapped. "Willie's no more crazy than I am. Maybe it's the world that's crazy. I don't know. I thought I did, but I don't." His blue eyes looked after the retreating figure. "Three hundred years of wounded pride," he mused. "Three centuries of hurt dignity. Going down the road marching. What would happen if we let them—no, it's Goddamned impossible. . . ."

"Looney!" Martha sobbed. "Plum tetched!"

"They'll kill him," Colonel Bob said. "And they'll do it in the meanest damned way they can think of. His leg won't make any difference. Not all the dang blasted ribbons in the world. Crazy thing. Willie, a soldier of the republic—wounded, and this thing to happen. Crazy." He stopped suddenly, his blue eyes widening in his pale, old face. "Crazy!" he roared. "That's it! If I can make them think—That's it, that's it, by God!"

Then he was racing through the house toward the telephone.

Willie had gone on around the house toward the dirt road, where the heat was a visible thing, and turned his face in the direction of town.

When he neared the one paved street, the heat was lessening. He walked very slowly, turning off the old country road into Lee Avenue, the main street of the town. Then he was moving toward the station. There were many people in the street, he noticed, far more than usual. The sidewalk was almost blocked with men with eyes of blue ice, and a long, slow slouch to their walk. He went on quietly, paying no attention to them. He walked in an absolutely straight line, turning neither to the right nor the left, and each time they opened up their ranks to let him pass through. But afterwards came the sound of their footsteps falling in behind him, each man that he passed swelling the number until the sound of them walking was loud in the silent street.

He did not look back. He limped on his artificial leg making a scraping rustle on the sidewalk, and behind him, steadily, beat upon beat, not in perfect time, a little ragged, moving slowly, steadily, no faster nor slower than he was going, the white men came. They went down the street until they had almost reached the station. Then, moving his lips in prayer that had no words, Willie turned and faced them. They swung out into a broad semicircle, without hastening their steps, moving in toward him in the thick hot silence.

Willie opened his mouth to shriek at them, curse them, goad them into haste, but before his voice could rush past his dried and thickened tongue, the stillness was split from top to bottom by the wail of a siren. They all turned then, looking down the road, to where the khaki-colored truck was pounding up a billowing wall of dust, hurling straight toward them.

Then it was upon them, screeching to a stop, the great red crosses gleaming on its sides. The two soldiers were out of it almost before it was still, grabbing Willie by the arms, dragging him toward the ambulance. Then the young officer with the single silver bar on his cap was climbing down, and with him an old man with white hair.

"This the man, Colonel?" the young officer demanded.

Colonel Bob nodded.

"All right," the officer said. "We'll take over now. This man is a combat fatigue case—not responsible for his actions."

"But I got to go!" Willie said. "Got to ketch that train. Got to go North where I can be free, where I can be a man. You hear me, lieutenant, I got to go!"

The young officer jerked his head in the direction of the ambulance.

"Let me go!" Willie wept. "Let me go!"

But the soldiers were moving forward now, dragging the slim form with them, with one leg sticking out very stiffly, the heavy heel drawing a line through the heat-softened asphalt as they went.

Frank Yerby

QUESTIONS: IDEAS TO THINK ABOUT

1. *How has the experience of blacks in the armed forces changed during this century? Are there examples of racism in the military today?*
2. *Compare the reception returning veterans received after Vietnam with World War II and World War I.*
3. *Do you feel the optimism stated in the final line of the poem, "Sonnet to Negro Soldiers," is justified in American society today?*
4. *In Frank Yerby's story, "The Homecoming," what is the irony of the inscription on the statue of the Confederate soldier? How has Willie been changed by his war experiences? What ideas have made Willie "sick" in the eyes of the white people in his home town?*

BLACK CAPITALISM

As we saw earlier when we looked at the Geipel family, owning one's own business, being one's own boss, is an integral part of the American dream. This has been true for black Americans as well as white, although the struggle by blacks to achieve a capitalist identity has certainly been more difficult. Many would-be capitalists encountered racism in addition to the usual problems of starting one's own business.

In a positive sense, capitalism means working for one's self. On the other hand, it also means that the capitalist makes a profit from the work of others. Throughout American history it has unfortunately been the latter definition that has been the experience of most workers. Even under slavery, blacks saw the economic realities of the system whereby owners grew rich from their slaves' labor. This understanding found expression in one of the slaves' work songs:

The big bee flies high,
The little bee makes the honey;
The black folks makes the cotton
And the white folks gets the money.

Lawrence Levine,
Black Culture and Black
Consciousness

IN BUSINESS

The following poem by Maya Angelou about a youngster working at a shoeshine stand, singing, "I'm a cap'tilist, can you dig it?" reflects the energy of capitalism that is basic to contemporary black consciousness.

TIMES SQUARE SHOESHINE COMPOSITION
BY MAYA ANGELOU

I'm the best that ever done it
(pow pow)
that's my title and I won it
(pow pow)
I ain't lying, I'm the best
(pow pow)
Come and put me to the test
(pow pow)

I'll clean 'em til they squeak
(pow pow)

In the middle of next week,
(pow pow)
I'll shine 'em til they whine
(pow pow)
Till they call me master mine
(pow pow)

For a quarter and a dime
(pow pow)
You can get the dee luxe shine
(pow pow)
Say you wanta pay a quarter?
(pow pow)
Then you give that to your daughter
(pow pow)

I ain't playing dozens' mister
(pow pow)
You can give it to your sister
(pow pow)
Any way you want to read it
(pow pow)
Maybe it's your momma need it.
(pow pow)

Say I'm like a greedy bigot,
(pow pow)
I'm a cap'tilist, can you dig it?
(pow pow)

IN BUSINESS. In this photograph George Aaron stands beaming beside his son's hot dog truck. After decades of struggle, of working for others, George and his family see in that truck much more than simply a way to earn a living; it represents the achievement of economic independence that escapes most people during their lifetimes.

THE MANYOKY FAMILY

Like the Geipels and the Bonner-Cannon-Aarons, the story of the Manyokys begins with migration to Queens. Emory Manyoky emigrated from Hungary in 1901 and did odd jobs in woodworking. He was a cabinet maker from the old country and, in his son's words, "saved up money from his measly wages to send for my mother, my sister and I to take a steamship to New York in 1903."

In 1909, Emory Manyoky went to work at the Steinway Piano Company. Ten years later his son, Jess, joined him at Steinway, and together they spent ninety years as craftsmen for the company.

The work world of the Geipels and the Bonner-Cannon-Aarons was shaped by their respective communities, and the Manyoky work world was also shaped by the same forces. The years of the Manyoky's employment saw important changes in the relations between workers and management, a shift from a paternalistic system in which bosses saw workers as part of a family to a relationship in which workers joined together to form a union.

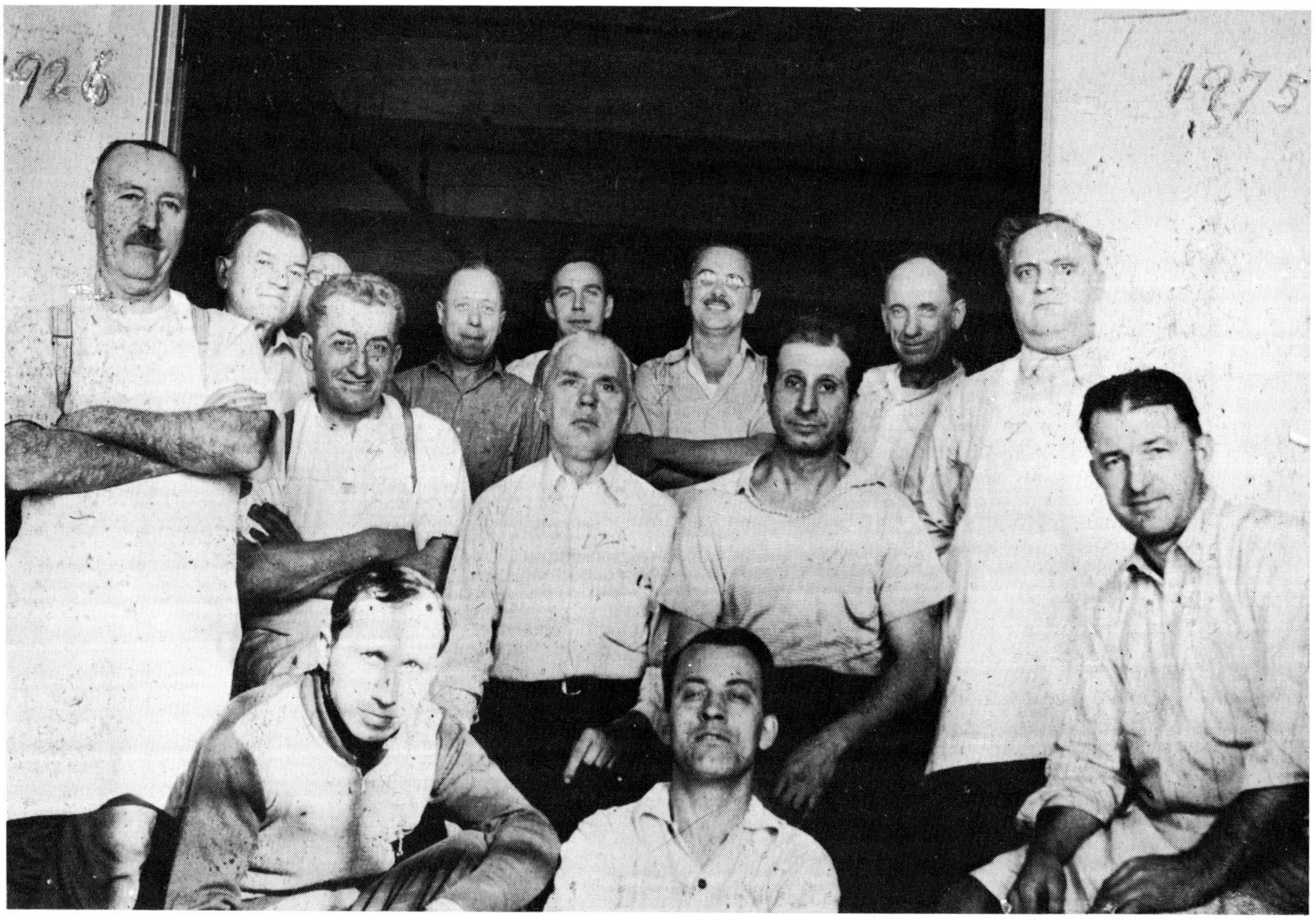

JESS MANYOKY AND HIS CO-WORKERS—1943. "I worked 53 years at Steinway. It was a family affair, from the porter to the superintendent." — Jess Manyoky.

JESS MANYOKY AND CO-WORKERS AT THE STEINWAY PIANO FACTORY, 1943

Jess Manyoky (center of foreground in photograph) worked in the grand action regulating department of the Steinway Piano Factory. The department members posed for the group picture reproduced here. It was Manyoky's job to shave the individual felt hammers which strike the keys. Through this "voicing" process, Manyoky and his co-workers could tone a piano by softening or firming up individual notes. Their skills turned an ordinary piano into a Steinway.

Jess Manyoky recalled his father's experience: "In 1919, my father [appointed himself] a delegate of his department—there was no union then, although there had been one in the past. Nine to ten men from different departments were fired for two years after trying to negotiate with the bosses and organizing a one week strike. My father was one of them. After World War I began, the bosses found they were stuck for European help—they need skilled help to get production out the way they used to. So they took a chance on a kid like me—19, with no woodworking experience." Manyoky attributes the opportunity to a good "connection": "I was a boy scout at that time, and the scoutmaster's father was superintendent of the plant—so that was an 'in' for me, to be a boy scout."

Emory Manyoky was called back to work the same year his son was hired—1921. The message came in a letter from Steinway management saying: "The war is over, we got work upstairs if you want to come back." The elder Manyoky trained his son in the grand action regulating department, teaching him the skills involved in regulating all the moving parts of the piano such as the foot pedal.

All together, the Manyokys spent 90 years at Steinway. Emory retired in 1946 after 37 years, and Jess retired in 1975 after 55 years.

QUESTIONS: IDEAS TO THINK ABOUT

1. *Why might this photograph have been taken? How does it compare to a family photograph?*
2. *Do you know any crafts people? How did they learn their crafts? From a parent?*
3. *How did your parents and grandparents learn their trade?*
4. *How does piece work compare with assembly line work?*
5. *What are the effects on workers who stay with one company for their entire life?*

THANKING MY MOTHER FOR PIANO LESSONS

by Diane Wakoski

Jess Manyoky played a vital role in the production of a fine piano. Hearing the tone, richness, and resonance of this superb instrument touched by the fingers of a great pianist is an experience treasured by many of us. Although most of us never reach the concert stage, many of us have childhood memories of the frustrations and joys of learning to play the piano. Such memories are recalled in *Thanking My Mother for Piano Lessons*, a poem which pays tribute to the financial sacrifices made by a working class mother to give her lonely daughter the companionship of piano lessons. Those sacrifices gave Jess and other piano workers their jobs.

"THANKING MY MOTHER FOR PIANO LESSONS"
BY DIANE WAKOSKI

The relief of putting your fingers on the keyboard,
as if you were walking on the beach
and found a diamond
as big as a shoe;

as if
you had just built a wooden table
and the smell of sawdust was in the air,
your hands dry and woody;

as if
you had eluded
the man in the dark hat who had been following you
all week;

the relief
of putting your fingers on the keyboard,
playing the chords of
Beethoven,
Bach,
Chopin
        in an afternoon when I had no one to talk to,
        when the magazine advertisement forms of soft sweaters
        and clean shining Republican middle-class hair
        walked into carpeted houses
        and left me alone
        with bare floors and a few books

I want to thank my mother
for working every day
in a drab office
in garages and water companies
cutting the cream out of her coffee at 40
to lose weight, her heavy body
writing its delicate bookkeeper's ledgers
alone, with no man to look at her face,
her body, her prematurely white hair
in love

I want to thank
my mother for working and always paying for
my piano lessons
before she paid the Bank of America loan
or bought the groceries
or had our old rattling Ford repaired.

. . .

But I played my way
on the old upright piano
obtained for $10,
played my way through fear,
through ugliness,
through growing up in a world of dime-store purchases,
and a desire to love
a loveless world.

I played my way through an ugly face
and lonely afternoons, days, evenings, nights,
mornings even, empty
as a rusty coffee can,
played my way through the rustles of spring
and wanted everything around me to shimmer like the narrow tide
on a flat beach at sunset in Southern California,
I played my way through
an empty father's hat in my mother's closet
and a bed she slept on only one side of,
never wrinkling an inch of
the other side,
waiting,
waiting,

. . .

I played my way from age to age,
but they all seemed ageless
or perhaps always
old and lonely,
wanting only one thing, surrounded by the dusty bitter-smelling
leaves of orange trees,
wanting only to be touched by a man who loved me,
who would be there every night
to put his large strong hand over my shoulder,
whose hips I would wake up against in the morning,
whose mustaches might brush a face asleep,
dreaming of pianos that made the sound of Mozart
and Schubert without demanding
that life suck everything
out of you each day,
without demanding the emptiness
of a timid little life.

I want to thank my mother
for letting me wake her up sometimes at 6 in the morning
when I practiced my lessons
and for making sure I had a piano
to lay my school books down on, every afternoon.
I haven't touched the piano in 10 years,

. . .

When I touch the man
I love,
I want to thank my mother for giving me
piano lessons
all those years,
keeping the memory of Beethoven,
a deaf tortured man,
in mind;
of the beauty that can come
from even an ugly
past.

From By Women: An Anthology,
edited by Marcia McClintock Folsom
and Linda Heinlien Kirschner

QUESTIONS: IDEAS TO THINK ABOUT

1. *Why do you think the mother in the poem made such sacrifices for her daughter's piano lessons?*
2. *Does it surprise you when the daughter says, "I haven't touched it in years"? Why?*
3. *Explain what you think the piano meant to the author of this poem.*

ANNUAL WAGES EARNED, 1928 TO 1942

In 1928, the year before the stock market crash and the beginning of the Great Depression, Jess Manyoky made $2,309.70, mostly through piece-work. This income compared favorably with wages paid in other cities around the country and was much higher than the average wage in Queens ($1,472). It put Jess in the highest class of factory workers and afforded him a fairly comfortable lifestyle.

During the early 1930's, however, during the worst years of the Depression, Steinway ran the factory only on a part-time basis; in 1932 it was shut down entirely. The effects can be seen in Jess' wages, which fell to a scant $140 and only slowly and erratically crept upwards again. Even in 1942, when America was on a wartime economy and industrial production was again rising, Jess' annual income was less than he had earned before the Depression.

ANNUAL WAGES EARNED—1928 TO 1942.

average for
11 yrs. total 11310.14 per yr 1028.19

1928	2309 70	1942	2165.26
1929	2429.16		
	1609.20		
1930	670 00		
1931	140 00		
1932	335 58		
1933	657 24		
1934	801 82		
1935	1601 83		
1936			
1937	2207 57		
1938	538.40		
1939	1561 85		
194[illegible]	1692.23		
	2199.18		

courtesy of Jess Manyoky.

BANK	CITY	AVERAGE WAGES PAID
1	Highland Park (Mich.)	$1,917.56
2	Hamtramick (Mich.)	1,797.43
3	Youngstown	1,752.91
4	Flint	1,712.05
5	Detroit	1,701.36
6	New York City	1,567.51
7	Toledo	1,544.00
8	Akron	1,531.52
9	Chicago	1,523.17
10	Cleveland	1,515.80
11	Bayonne (N. J.)	1,500.38
12	Dayton	1,488.19
13	Los Angeles	1,477.54
14	Queens	1,472.35
15	San Francisco	1,462.92
16	Pittsburgh	1,444.00
17	Buffalo	1,440.58
18	Rochester	1,395.84
19	Jersey City	1,388.45
20	Omaha	1,361.19
21	Milwaukee	1,357.62
22	Philadelphia	1,347.55
23	Boston	1,342.38
24	Newark	1,340.83
25	Paterson (N. J.)	1,332.29
26	St. Paul	1,297.38
27	Kansas City (Mo.)	1,285.74
28	Cincinnati	1,271.22
29	Minneapolis	1,270.43
30	Perth Amboy (N. J.)	1,258.90
31	St. Louis	1,245.99
32	Kansas City (Kan.)	1,243.21
33	Indianapolis	1,229.38
34	Providence	1,206.35
35	Louisville	1,150.86
36	Baltimore	1,091.08

	PER YEAR
Manhattan	$1,642
Richmond	1,528
Bronx	1,507
Queens	1,472
Brooklyn	1,437

QUESTIONS: IDEAS TO THINK ABOUT

1. *What happened to Jess during the Depression?*
2. *According to this document, what was the worst year for Jess? How do you think that affected him?*
3. *What happened to your family during the Great Depression?*
4. *What happens to a family when the breadwinner is unemployed for a year or more?*

A & P FOOD STORES ADVERTISEMENT, 1933

This advertisement from the Long Island Star lists prices in 1933, in the midst of the Depression. Prices of most items are approximately one-tenth of what the same items cost today. It must be remembered, however, that in that year Jess Manyoky earned only $335.18, an average of less than $7 a week.

The eagle in the upper right hand corner is the symbol of the National Recovery Administration (NRA). The National Industrial Recovery Act set up the NRA, which administered codes that set prices, production quotas, and wages scales. Although the act was later declared unconstitutional, it had demonstrated that President Franklin D. Roosevelt's new administration was attempting to address the country's economic problems.

A & P FOOD STORES ADVERTISEMENT. *The Daily Star, 1933*

A & P AD—Daily News, April 14, 1982

QUESTIONS: IDEAS TO THINK ABOUT

1. *In 1933, the date of this ad, Jess earned less than $7 a week. What could he afford to buy on that amount?*
2. *How do the prices in the 1933 ad compare with the prices in the ad from 1982? For example, in 1933 macaroni cost 17¢ for three 8 oz. packages; in 1982, Ronzoni noodles cost 59¢ for one 12 oz. package.*
3. *Has the ratio between food costs and salaries changed between 1933 and 1982?*

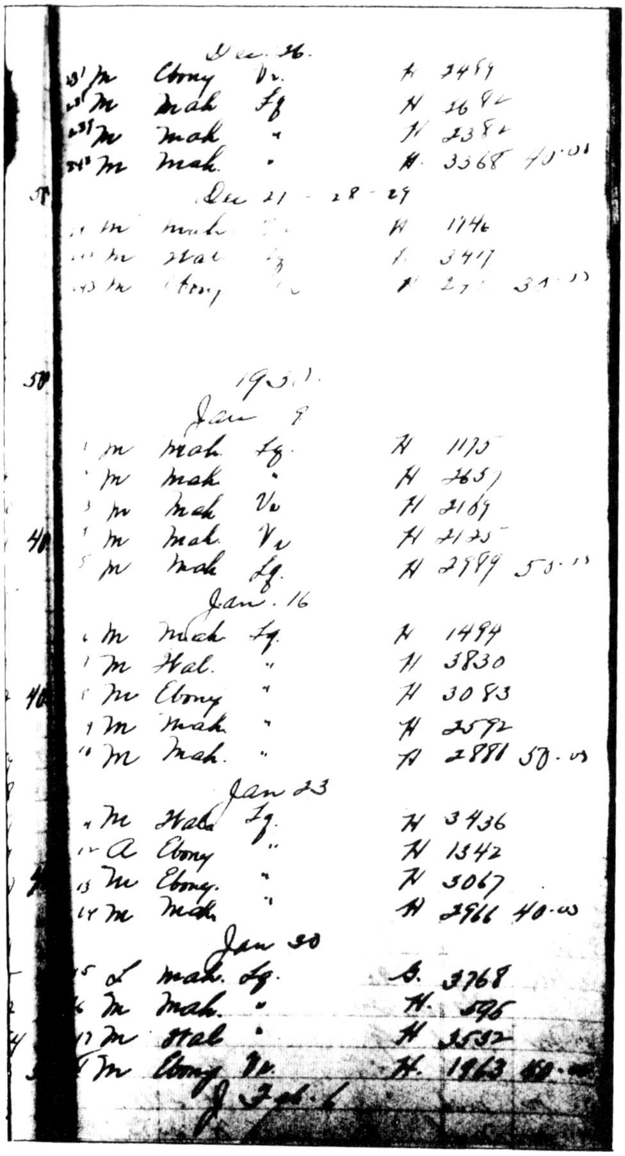

PAY RECORD—DECEMBER 26, 1929 TO FEBRUARY 6, 1930. Jess Manyoky kept meticulous records of his work production and salary at Steinway. According to this document, Manoky "tone regulated" 243 pianos in 1929. When he first joined Steinway in 1921, Manyoky made $11.93 a week. By 1930, he made $40-$50 a week. The notations in the ledger indicate the model number of the piano and whether it was mahogany, walnut or ebony.

QUESTIONS: IDEAS TO THINK ABOUT

1. According to this document, how much did Jess earn for each piano he regulated?
2. Like most of the workers at Steinway, Jess was paid on the basis of piece-work. What are the advantages and disadvantages compared with being paid by the hour? Which system is more advantageous for the worker? for management?
3. How did Manyoky's work compare to Nancy Aaron's work in a shoe factory or Wilhelmine Geipel's work as a midwife?
4. Do you save your work records? Do you have work records belonging to your parents or grandparents?

WORKING GIRLS' BUDGETS

by Sue Ainslie Clark and Edith Wyatt, 1910

When we look at the pay records of Jess Manyoky, we learn a great deal about salaries and the difficulty of making ends meet. We also saw what happened to Jess' salary when the Depression hit. Nevertheless, Jess was much better off than most other workers, particularly unmarried, unskilled female factory workers.

Sue Ainslie Clark and Edith Wyatt wrote *Working Girls' Budgets* in 1910 for *McClure's Magazine*. The article describes the working conditions of women workers in New York City, using the personal work histories of several immigrant women. One of them earned only $239.50 a year; another made between $4 and $6 a week. (Nancy Aaron, you remember, only made $5 a week when she worked in the shoe factory in 1914.)

WORKING-GIRLS' BUDGETS

III

Unskilled And Seasonal Factory Workers

Among the women workers who help to supply the enormous garment trade of New York with the innumberable yards of stitching it requires, the waist-makers have within the last year achieved, through organization, a better spirit of solidarity and hours less exhausting.

What are the chances in life of some of the other self-supporting girls and women who are working in factories, in the garment trade and other industries in New York? Besides the accounts of the waist-makers, the National Consumers' League received, in its inquiry on this point, specific chronicles from skilled and from unskilled factory workers, both hand workers and machine operatives—among others, packers of drugs, biscuits, and olives, cigarette-rollers, box-makers, umbrella-makers, hat-makers, glove-makers, fur-sewers, hand embroiderers, white-goods workers, shirt-makers, workers on men's coats, and workers on children's dresses.

Experiences of Yeddie Bruker, a Hungarian White-Goods Worker

The chronicles printed below, taken from establishments of different kinds and grades, express as clearly as possible the several features most common to the trade fortunes the workers described—uncertain and seasonal employment, small exploitations, monotony in occupation, and fatigue from speeding.

Because of uncertain and seasonal employment, machine operatives in the New York sewing industries frequently change from one trade to another. This had been the experience of Yeddie Bruker, a young Hungarian white-goods worker living in the Bronx.

The tenements of the Bronx appear as crowded as those of the longer-settled neighborhoods of Manhattan, the lower East Side, Harlem, Chelsea, and the cross streets off the Bowery, where so many self-supporting factory workers live. These newer-built lodgings, too, have close, stifling halls, and inner courts hung thick with washing. Here, too, you see, through the windows, flower-makers and human hair workers at their tasks; and in the entries, hung with Hungarian and German signs, the children sit crowded among large women with many puffs of hair and striking preference for frail light pink and blue princess dresses. These blocks of Rumanian and Hungarian tenement districts, their fire-escapes hung with feather-beds and old carpets, and looking like great overflowing waste-baskets, are scattered in among little bluff ledges, scraggy with walnut brush, some great rocks still unblasted, and several patches of Indian corn in sloping hillside empty lots—small, strange heights of old New York country, still unsubmerged by the wide tide of Slav and Austrian immigration.

In this curious and bizarre neighborhood, Yeddie Bruker and her sister lived in a filthy tenement building, in one room of an extremely clean little flat owned by a family of their own nationality.

Yeddie was a spirited, handsome girl of twenty-one, though rather worn-looking and white. At work for six years in New York, she had at first been a machine operative in a large pencil factory, where she fastened to the ends of the pencils the little corrugated tin bands to which erasers are attached. Then she had been a belt-maker, then a stitcher on men's collars, and during the last four years a white-goods worker.

Mrs. Hallett lived in an excessively small, unheated hall bedroom, on the fourth floor of an enormous old house filled with the clatter of the elevated railroad. On the night of the inquirer's call, she was pathetically concerned lest her visitor should catch cold because "she wasn't used to it." She lighted a small candle to show her the room, furnished with one straight hard chair, a cot, and a washstand with a broken pitcher, but with barely space besides for Mrs. Clark and her kind, public-spirited little hostess. They sat, drowned at times in the noise of the elevated, in almost complete darkness, as Mrs. Hallett insisted on making vain effort to extract some heat for her guest from the single gas-jet, by attaching to it an extremely small gas-stove.

For this room, which was within walking distance of the candy factory, Mrs. Hallett paid $1.75 a week. Her breakfast of coffee and rolls in a bakery near by cost her 10 cents daily. She apportioned 15 or 25 cents each for her luncheon or dinner at restaurants. In her hungriest and most extravagant moments she lunched for 30 cents. Her allowance for food had to be meager, because, as she had no laundry facilities, she was obliged to have her washing done outside. Sometimes she contrived to save a dollar a week toward buying clothing. But this meant living less tidily by having less washing done, or going hungrier. During the last year her expense for clothing had been a little more than $23; summer hat, $1; winter hat, $1.98; best hat, $2; shoes (2 pairs at $2.98, 2 pairs rubbers), $7.16; wrap (long coat), $2.98; skirt (a best black brilliantine, worn two years), at $5.50, $2.75; underskirt (black sateen), 98 cents; shirtwaist (black cotton, worn every day in the year), 98 cents; 2 union suits at $1.25 (one every other year), $1.25; 6 pairs stockings at 25 cents, $1.50; total $23.56.

She said with deprecation that she sometimes went to the theater with some young girl friends, paying 25 cents for a seat, "because I like a good time now and then."

These trade fortunes represent as clearly as possible the usual industrial experience of the women workers in unskilled factory labor who gave accounts of their income and outlay in their work away from home in New York.

Sue Ainslie Clark and Edith Wyatt

QUESTIONS: IDEAS TO THINK ABOUT

1. *After reading this article, what details stand out in your mind? Does the article shock you? Why?*
2. *The authors of this article, Sue Ainslie Clark and Edith Wyatt, were in the muckraking tradition of Upton Sinclair, exposing the underside of life in America.*
3. *Compare Nancy Aaron's story with the stories of the women in this article.*

ORAL HISTORY INTERVIEW WITH JOHN STEINWAY

We have looked at work at Steinway and Sons from the perspective of the workers, but that is not the complete story. We must also hear from the Steinways themselves, a family of German immigrants who built the company from their own resources. A theme that runs through this interview is that the company was like a family.

It is interesting to note that John Steinway himself worked in the factory as an apprentice. The man who would later own the company learned the business from the workers as well as his father.

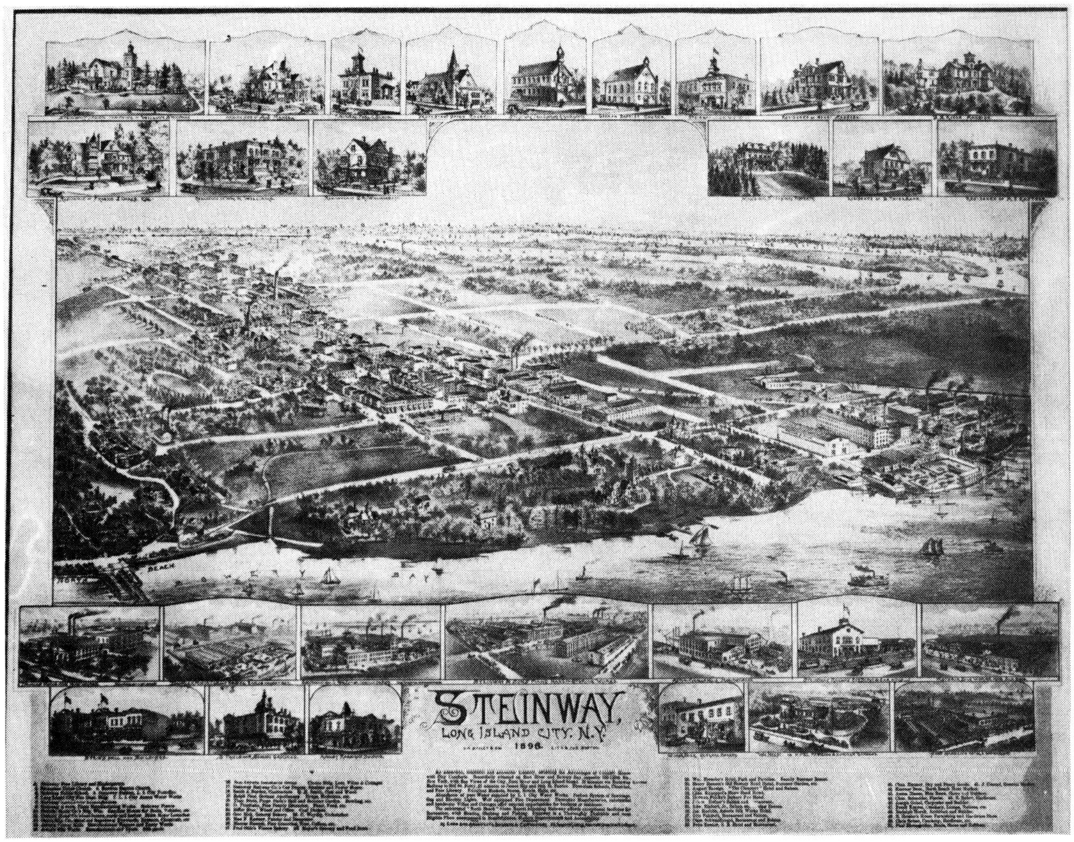

QUESTIONS: IDEAS TO THINK ABOUT

1. *From this interview, what do you think John Steinway's attitude toward his employees is? Can you find specific quotes that gave you that impression?*
2. *What does John Steinway want you to know about his family's success story?*
3. *Knowing what you do about Jess Manyoky's experience at Steinway, how do you think he would respond to this interview?*
4. *Manyoky and Steinway experienced America in quite different ways, despite their connection at the company. Would either of them say they had realized the "American Dream"? Would they measure success in the same way?*

AN ORAL HISTORY INTERVIEW WITH JOHN STEINWAY

We have looked at the Steinway factory through the eyes of a worker; now let us look at it through the eyes of a Steinway:

My great grandfather, Henry Englehart Steinway, came to this country in 1850 after the Revolution of 1848 in Central Germany. He was a native of the Hartz Mountains in the Duchy of Brunswick.

Henry Steinway was an old man of 53 when he came over here with a wife and seven children. It's interesting to note that he sent one of his sons over here a year before to Havana and to New York. We have in the family archives—some letters from Charles Steinway back to his papa in Seesen, in Brunswick, in German, describing New York in 1849 which are perfectly fascinating. They describe this country, the opportunity to make a living over here, the freedom of action, and of thought over here as opposed to Central Germany in the mid 19th Century. It was "O.K." he wrote, to pack up mom and pop and the kids and come on over, and they did.

They came here on June 9, 1850, and they came to Hester Street which is the street just north of Canal. They lived at 199 Hester Street for three years; obviously immigrant people who didn't speak much English. They learned about this country, and two of them lost their jobs. The obvious decision was, "Look, if we're going to starve, we might as well starve together." So they started their own little business called Steinway and Sons. It was the old man and his three sons.

Being methodical Germans, they kept beautiful records, and we have the business records today. The earliest records are kept in the handwriting of the fourth son, a boy of 13 when he came over. He was obviously the first one to read and write English because they said "William you make with the English, you keep the records." The records are kept laboriously in his handwriting. He happened to be my grandfather.

They worked hard, the business prospered, and very soon began to expand. They started on Walker Street, the street just south of Canal Street in Manhattan. They rented a couple of buildings, lived upstairs, and made the pianos downstairs. They made square pianos in those days, what is known in German as the "Tafelklavier" a very old-fashioned type of piano but fashionable in the mid 19th Century. Eighty-two and Eighty-four Walker Street was the site of the first Steinway factory.

When they got successful, they moved up to 14th Street and built a building at 109 E. 14th Street. The opening of the building was May 28, 1865. In 1865, they built a fancy showroom here on 14th Street which was Steinway Hall until we sold it to Mr. S. Klein, who had a little dress shop in that area.

Pretty soon with prosperity, Steinway and Sons decided they were going to move way uptown. So they bought a piece of property and they built a five-story factory building on the corner of Park Avenue and 53rd Street. Just an interesting sidelight: when they were building the Seagrams building on that site, they had one of those wood railings around the portholes so you could watch what the boys were doing down in the hole. I went by one day, and a little loudspeaker and tape announced to all the people, "We're building this beautiful Seagram's building, etc. and it was once the site of the Steinway piano factory." Well, I was interested in that so I went around to the construction shack and asked for the boss. This big burly Irishman in the back of the office said, "What's your name?" and I told him. He said "you come with me." I went out of the shack and we went down one of those ladders and over down to the bottom of the hole. They were digging foundations. There were about eight guys with their hammers banging away at the foundation. He turned to me and said, "Look"—and there was a fieldstone and mortar foundation 12 feet thick. It was the foundation of the old factory. He said, "Why did you build a foundation 12 feet thick for a five-story factory building?" I replied, "They were Germans; they wanted it built solid."

My family moved to Astoria, Long Island in 1871, and after they sold the 53rd Street factory in 1906, they finally moved the whole operation out there. That was the Ditmars factory, but the Riker factory or the 19th Avenue factory where we are today was an original purchase of seven farms in 1871, including the estate, which is the old mansion.

The old mansion which still stands was built by Benjamin F. Pike. Pike was the great eyeglasses expert in the City of New York and he made a great deal of money making glasses, and he built this mansion in the 1850's. After he died, his widow and family sold the estate in July, 1870 to the Steinway family, my grandfather and his brothers. The seven farms constituted 440 acres, all told, and is pretty much the land between Bowery Bay and the present Grand Central Parkway.

The first Mercedes Benz made in America was made right on Steinway Street. Yes sir, the side line of my grandfather's. Gottlieb Daimler of Germany assigned to William Steinway of New York all the American marks and German patents for Daimler's Mercedes-Benz automobiles, to William Steinway, dated 1888 in September. In Montvale, New Jersey where Mercedes-Benz has its North American headquarters, they have in the front lobby of that building a Mercedes-Benz built in Astoria, in 1905, in the factory which was halfway between Shore Road and Old Shore Road on Steinway Street.

We had some wonderful characters in that business over the years. Charlie Braun was on the job 72 years in the factory. When I was an apprentice boy, I sat with him one day, having sandwiches or something, and I said, "Charlie, how did you get your job?" So Charlie said, "My father brought me down to the 53rd Street factory when I was 12 years old to go to work. He brought me to your grandfather and said 'Guten Morgen, Herr Steinway, my son here Charlie big boy, time he went to work.' And Mr. Steinway got over the desk and looked down at me and said, 'Braun take him home and fatten him up, for a year."

So Charlie said for his 13th birthday his papa took him down to the factory, and said "Herr Steinway, I fattened him up" and he started working, and 72 years later he was still there, on the job. A wonderful character, an absolutely sweet charming old fellow. That's what makes Astoria what it is.

A WORKER READS HISTORY
by Bertolt Brecht

Who built the seven towers of Thebes?
The books are filled with names of kings.
Was it kings who hauled the craggy blocks of stone?...
In the evening when the Chinese wall was finished
Where did the masons go?...

Young Alexander plundered India.
He alone?
Caesar beat the Gauls.
Was there not even a cook in his army?
Philip of Spain wept as his fleet
Was sunk and destroyed. Were there no other tears?
Frederick the Great triumphed in the Seven
Years War. Who
Triumphed with him?...

Every ten years a great man,
Who paid the piper?

So many particulars.
So many questions.

—Bertolt Brecht

QUESTIONS: IDEAS TO THINK ABOUT

1. *What does this poem have to say about the relationships between "Great Men" and workers?*
2. *If you could go back in time, what questions would you ask of the masons who built the Great Wall of China?*

LEISURE

INVITATION TO SEEING IN PERSPECTIVE

In the history of North Beach, an area that Queens residents used for recreation, we find a sample of the history of leisure in our culture. The North Beach story raises questions of how leisure became a business, how much leisure did people have then and how much now, and how you decide which are public and which are private spaces. Also underlying this history are the issues of a city growing under the pressures of race and class and rapidly changing technology.

As New York City grew between the end of the Civil War (1865) and the end of the First World War (1918), the populations changed dramatically, and their ideas of how to spend their spare time are a reflection of changes in attitudes and environment.

We must look at a number of trends to understand this process. First, we must ask what people can afford and if what is available to them suits their needs. Next, we must understand that the middle and upper class looked at some working class pleasures as morally wrong. Between 1865-1918, hundreds of thousands of European immigrants entered the city. We must explore if leisure patterns from the Old World lasted into the New World, into New York, and whether North Beach appealed to those tastes and attitudes. Finally, we must understand the effect of electricity, motion picture, and recorded music. Each technological advance had an effect on the way individuals spent their leisure time.

Exploring the history of North Beach shows the changing pattern of leisure over time. Through photographs, oral testimony, postcards and other printed materials, and through works of writers and artists, we will explore this changing pattern of leisure.

THE OPEN LAND, 1869-1905

In the mid-seventeenth century, Europeans settled the area along the shore of Bowery Bay, but for more than two hundred years it remained sparsely inhabited. A few farmhouses dotted the landscape, and the area produced fruits and vegetables for New York's growing population until the twentieth century. During the second half of the nineteenth century people came from Manhattan and Brooklyn as well as Queens to spend their leisure time along this shore. They swam in the clear water, picnicked under broad oaks, and breathed the fresh salt air. Fishing and crabbing drew the fishermen, and in autumn, when the nearby wetlands attracted ducks and geese on their annual migration to the south, the hunters came to shoot. Indeed, the nature of the landscape determined the kinds of activities people enjoyed there.

The joys of nature have stimulated people in all ages, and to capture in words the serenity of a scene beyond the city writers have given their best efforts. Notice how both John Keats and Henry David Thoreau acknowledge the soothing land of the outdoors.

To One Who Has Been Long in City Pent

To one who has been long in city pent
is very sweet to look into the fair
And open face of heaven,—to breathe
a prayer
Full in the smile of the blue firmament.
Who is more happy, when, with hearts content,
Fatigued he sinks into some pleasant lair of wavy
grass and reads some debonair
And gentle tale of love and languishment?
Returning home at evening, with an ear
Catching the notes of Philomel,—an eye
Watching the sailing cloudlet's bright career,
He mourns that day so soon has glided by:
the passage of an angel's tear
That falls through the clear ether silently.

John Keats, 1816

SOLITUDE

This is a delicious evening, when the whole body is one sense, and imbibes delight through every pore. I go and come with a strange liberty in Nature, a part of herself. As I walk along the stony shore of the pond in my shirt-sleeves, though it is cool as well as cloudy and windy, and I see nothing special to attract me, all the elements are unusually congenial to me. The bullfrogs trump to usher in the night, and the note of the whip-poor-will is borne on the rippling wind from over the water. Sympathy with the fluttering alder and poplar leaves almost takes away my breath; yet, like the lake, my serenity is rippled but not ruffled. These small waves raised by the evening wind are as remote from storm as the smooth reflecting surface. Though it is now dark, the wind still blows and roars in the wood, the waves still dash, and some creatures lull the rest with their notes. The repose is never complete. The wildest animals do not repose, but seek their prey now; the fox, and skunk, and rabbit, now roam the fields and woods without fear. They are Nature's watchmen,—links which connect the days of animated life.

Henry David Thoreau

We often forget that little more than a hundred years ago, people living in or visiting the North Beach area had easy access to the shore line, the woods, animals, silence. Looking out over the water a watcher could have easily imagined voyages on large sailing ships into the realms of great whales, or turning inland like Thoreau, heard the twilight sounds of frogs and rustling of leaves.

QUESTIONS: IDEAS TO THINK ABOUT

1. *Where in your neighborhood can you go for an evening's "shirt sleeve" walk?*
2. *What impact has crime and overcrowding in our cities had on our capacity to appreciate an experience such as Thoreau described in "Solitude"?*

PANORAMA OF BOWERY BAY, 1869

PANORAMA OF BOWERY BAY, 1869. The photograph, commissioned by Henry Steinway in preparation for relocating his piano company from Manhattan to Queens, along the bank of the East River, shows a panorama of Bowery Bay. In the front is Old Bowery Bay Road, which ran near the shoreline from Astoria to Flushing. Bowery Bay is on the far right of the panorama. That land became North Beach. The land in the far distance is College Point.

OARSMEN AT CHATOU
by Auguste Renoir, 1879

In this picture the French Impressionist artist Auguste Renoir paints bourgeois (middle-class) French enjoying the banks of the River Seine. We can picture similar scenes taking pace on the shore of Bowery Bay around the turn of the century, with the important difference that the members of the boat clubs in Queens were workers, not members of the middle-class. The individuals in the painting enjoy a day in the bright sun and away from the city.

REGATTA ON FLUSHING BAY, 1901

Compare Renoir's bright painting with this photograph of a regatta on Flushing Bay in 1901. Both attest to the popularity of boating, at least among those who could afford equipment and had access to the river or bay. In the late nineteenth century numerous clubs had their boathouses along the Queens shoreline from Astoria to College Point. These clubs raced against others from Manhattan, Brooklyn, the Bronx, and Connecticut. Judging by the crowd on the pier the events were popular indeed.

"BOWERY BAY BATHING"

from the *Long Island Star, August 7, 1885*

Although the activities described seem quite harmless to us today, the tone of this 1885 article expresses moral outrage at the activities along the shore of Bowery Bay. The story implies another issue—the ongoing dispute between private rights and public access. The author quite passionately describes the plight of residents in the area.

Bowery Bay Bathing

A visitor to the beach at Bowery Bay last Sunday afternoon in the vicinity of the mansion of the late Dr. Douglas and near the estates of Mr. Rapalyea and Mr. D. S. Riker would have imagined himself at Coney Island or Rockaway, minus, however, the costumes on the forms of the bathers that are required at those places. From midday until late in the afternoon wagon after wagon drove up to the beach containing parties of men, women and half-grown boys and children from all over New York and Brooklyn and in a short time, the beach was crowded with human forms darting hither, sometimes into the waters of the bay and then back into the grounds surrounding the houses in the neighborhood.

At some parts of the beach picnic parties were grouped beneath the trees and in some cases the women ever and anon ran bare-legged into the water and catching hold of some refractory "bad boy" or thoughtless girl, administered that maternal correction which is usually supposed to be confined exclusively to the woodshed or privacy of the family bedchamber. On these occasions shouts of laughter from the onlookers would ring out upon the air and the discomfited youth would retire with all haste to the shelter of the bushes while the maternal corrector would wage a war of angry vituperation with the amused audience.

The place where these scenes have been repeated on nearly every fine Sunday during the present summer is at the boundary line between Long Island City and Newtown. Only a few days ago a gentleman was stopped and robbed by a crowd of roughs as he passed the spot but although the police of this city and Newtown have made every effort to arrest the offenders, they have been so far unsuccessful. The indecent manner in which bathing is conducted is however apparent every week and can be and should be promptly suppressed. The indiscriminate mingling of the sexes, the brevity of covering and in some cases, the entire absence of any on the part of some of the men and women is indelicate in the extreme and will be likely, unless repressive measures are at once instituted, to lead to disorders of the worst character.

August 7, 1885 2:3

QUESTIONS: IDEAS TO THINK ABOUT

1. *The rights of the public to access to the shoreline is still a controversial issue. If you owned seashore property, would you wish to keep your beach private, or would you welcome city dwellers to share your space?*
2. *Our author in 1885 was concerned about the "proper" attire for swimming. Do we still have similar concerns today?*
3. *After reading this article, how true is the expression "the good old days"? Think about it in terms of violence and morality.*
4. *Can you find evidence of sexist attitudes in this article?*
5. *If you compare your grandparents, parents, and yourself, how has the idea of "a day at the beach" changed?*

ELIZABETH AUGENTE SAVED FROM DROWNING

Although the North Beach shoreline offered a pleasant and relaxing ambiance, there were also dangers. Drownings were all too common as swirling currents swept swimmers away. Sometimes children who played near the water without any supervision lost their footing, panicked, and drowned.

Here is an invitation to a party held by Elizabeth Augente's parents to celebrate "her narrow escape from drowning."

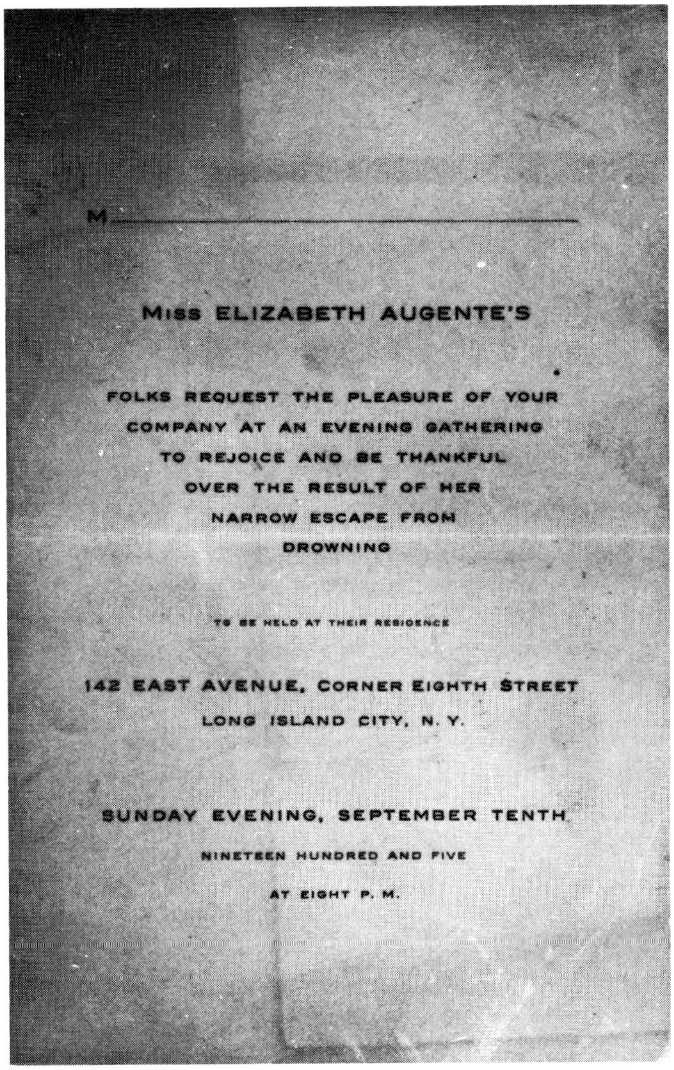

M______________________

MISS ELIZABETH AUGENTE'S

FOLKS REQUEST THE PLEASURE OF YOUR COMPANY AT AN EVENING GATHERING TO REJOICE AND BE THANKFUL OVER THE RESULT OF HER NARROW ESCAPE FROM DROWNING

TO BE HELD AT THEIR RESIDENCE

142 EAST AVENUE, CORNER EIGHTH STREET
LONG ISLAND CITY, N. Y.

SUNDAY EVENING, SEPTEMBER TENTH
NINETEEN HUNDRED AND FIVE
AT EIGHT P. M.

ELIZABETH AUGENTE SAVED FROM DROWNING, 1905.

QUESTIONS: IDEAS TO THINK ABOUT

1. *Recount an experience you or someone you know has had that was as potentially dangerous as Elizabeth's.*
2. *Imagine the East River, Flushing Bay and Bowery Bay to be clean and safe for swimming during the summer months. How would local swimming holes change the recreational patterns of today's New York?*

NORTH BEACH, 1889-1905

The atmosphere and the location, close to the city, attracted people to Bowery Bay. While some critics complained about immorality and lawlessness, others saw possibilities for profit.

George Ehret, owner of one of New York's largest breweries, and William Steinway, the piano manufacturer and owner of a street railway in Astoria, together with several other investors, formed the Bowery Bay Beach Improvement Company in late 1885. With a capital stock of $20,000 the company put their offices above the showroom of the Steinway Piano Company on 14th Street so they could plan a resort. Bowery Bay would be the worker resort for Steinway's "company town."

On opening day, June 19, 1886, more than 6,000 persons came to the new center. People crowded the hotel, bath house, picnic park, and dancing pavilion, taking advantage of the new and immediately popular leisure spot.

In 1890 the name Bowery Bay became North Beach; the company didn't want the resort area to be confused with the notorious Bowery in New York. In developing North Beach, the company kept two ideas in mind. First, they wanted to build a modern, attractive resort that was as safe and as clean as possible. They also wanted to please the tastes of their customers and to offer what these people expected. Steinway and Ehret themselves were German, therefore the resort had a decidedly northern European atmosphere. The food reflected these tastes: beer was the major drink, and the design carried the German influence.

This phase of development brought about a big change in the area, more than just building hotels and other facilities. The company changed the public space, areas where people could do as they please, to a private space where a commercial organization controlled many leisure activities and provided activities likely to bring profits.

HOTELS AND DANCE HALLS

Buildings owned by the Bowery Bay Beach Improvement Company housed nearly all the hotels and bars. For the first season the company ran each establishment directly. The next year, the operators leased the wooden open-air structures with the condition that they would sell only Ehret's beer. In later years, however, other breweries had saloons there. It was common in New York at that time for breweries to own saloons but for private individuals to run them. Ehret also owned numerous tenements scattered throughout the city, and when a tenant had trouble paying his rent, Ehret's agent often offered the man the chance to operate a saloon on the ground floor of one of Ehret's buildings.

The establishment at North Beach operated on a bigger scale than ordinary saloons and offered more than just beer and whiskey. Here one could find food, sausages, soup, and sandwiches. Usually there was entertainment, singers from New York or local talent. Customers could dance or listen to the live orchestra.

TIVOLI HOTEL, "SANDFORD'S POINT," NORTH BEACH, c. 1891.

The postcard shows Sandford's Point, a bit of land jutting into the bay. Here an old mansion bought by the Beach Company after the first season was transformed into a hotel. In the foreground is a cluster of picnic tables, evidently for families and "basket parties." Inside, the most important feature of the Tivoli House was the large airy dance hall.

Some people preferred to go inside the hotel to eat and drink, and others sat outside at the picnic tables. Some families may not have wanted to bring small children into the noisy, smoke-filled dance hall. The two spaces, inside and outside, also represented a class distinction; the patrons filling the hotel probably had more money than those who stayed outside.

HENRY DAUFKIRCH'S BAY VIEW HOTEL, 1905.

Henry Daufkirch's Bay View Hotel was one of the largest establishments at North Beach, offering "first-class accommodations" and facilities built to attract middle-class patrons. But the hotel also catered to workers' trolley parties. This advertisement came from the *North Beach Amusement Guide*, published in 1905 by the Beach Company.

QUESTIONS: IDEAS TO THINK ABOUT

1. *According to the ad, how could you reach the beach?*
2. *What can you figure out about the telephone number #43 B Astoria?*
3. *What ways does the ad use to attract people of different social classes?*
4. *If North Beach, as part of Queens, became incorporated into New York City in 1898, why do you think Daufkirch continued to advertise his hotel as in "Long Island"?*

MAGGIE, A GIRL OF THE STREETS
by Stephen Crane, 1983

In the selection below, Stephen Crane gives us a fine description of a local saloon in the Bowery at the turn of the century, an interior, probably very similar to Daufkirch's Bay View Hotel.

MAGGIE, A GIRL OF THE STREETS

An orchestra of yellow-silk women and bald-headed men, on an elevated stage near the center of a great, green-hued ball, played a popular waltz. The place was crowded with people grouped about little tables. A battalion of waiters slid among the throng, carrying trays of beer glasses, and making change from the inexhaustible vaults of their trousers pockets. Little boys, in the costumes of French chefs, paraded up and down the irregular aisles vending fancy cakes. There was a low rumble of conversation and a subdued clinking of glasses. Clouds of tobacco smoke rolled and wavered high in air above the dull gilt of the chandeliers.

The vast crowd had an air throughout of having just quitted labor. Men with callused hands, and attired in garments that showed the wear of an endless drudging for a living, smoked their pipes contentedly and spent five, ten, or perhaps fifteen cents for beer. There was a mere sprinkling of men who smoked cigars purchased elsewhere. The great body of the crowd was composed of people who showed that all day they strove with their hands. Quiet Germans, with maybe their wives and two or three children, sat listening to the music, with the expressions of happy cows. An occasional party of sailors from a warship, their faces pictures of sturdy health, spent the earlier hours of the evening at the small, round tables. Very infrequent tipsy men, swollen with the value of their opinions, engaged their companions in earnest and confidential conversation. In the balcony, and here and there below, shone the impassive faces of women. The nationalities of the Bowery beamed upon the stage from all directions.

Stephen Crane (1893)

QUESTIONS: IDEAS TO THINK ABOUT

1. *How do you know that the inhabitants of Crane's dance-hall are mostly working class? Does Crane suggest another class of people frequenting the saloon?*
2. *Why does Crane describe the German couples as "happy cows"? Do you find this description offensive? Why or why not?*
3. *Why are most of the women either with a husband, in the balcony or sprinkled "here and there" on the dance floor"?*

FRED. DEUTSCHMANN'S SILVER SPRING BATHING PAVILION

One of the main attractions of North Beach was the Bowery Bay. Instead of letting visitors swim along the open shoreline, however, the Beach Company built new bathing facilities. According to the *Long Island Star* of June 11, 1886, a week before the resort opened, the company built 130 bath houses, "fitted up with every modern convenience and divided into two sections, one being reserved for ladies and the other for gentlemen. By this arrangement the disgraceful scenes enacted on the beach last summer cannot possibly occur" (see page). Profit was only one reason for building these bath houses. Another motive was to force a standard of behavior and appearance on the patrons, and to keep out those who didn't live up to that standard.

The 1897 season ticket bought by the Geipel family suggests that mostly middle-class Germans patronized Silver Spring (both Geipel and Deutschmann were German). Since the pass was not good on Sundays or holidays, the Geipels must have gone to Silver Spring on weekdays.

For the majority of the workers in New York during the 1890's, the $4.00 cost of a season ticket at Silver Spring represented almost a week's wages. For most workers, a trip to North Beach was a special occasion, and they could not afford paying that much for a family outing more than once or twice a summer.

Dora Geipel, the woman on the left, was 23 years old when this picture was taken in 1918. She had been going to North Beach practically all her life, first with her family and later with her male and female friends. In the background is the restaurant and dance hall that was part of Silver Spring.

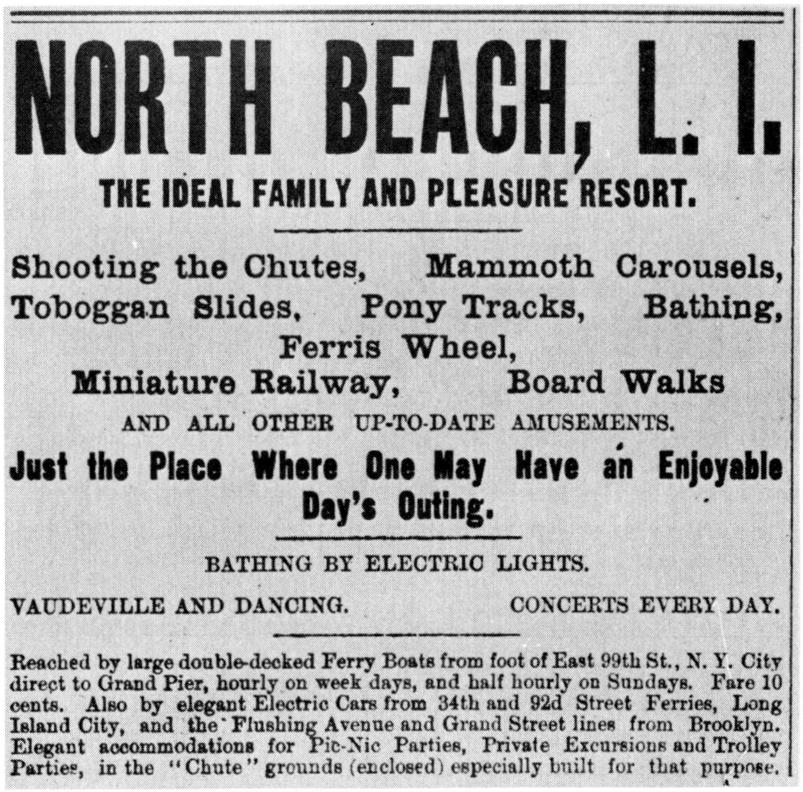

ADVERTISEMENT FOR NORTH BEACH, c. 1910.

DECLINE OF NORTH BEACH, 1892-1934

By 1890, when the Beach Company changed the name of the extremely popular resort from Bowery Bay to North Beach, several rides and attractions for all ages and pocketbooks were at the beach. Now, instead of attracting crowds with the ocean and the country atmosphere, the resort turned into a honky-tonk, and people came for thrills and excitement.

The changing tastes of the public, competition from the much larger parks at Coney Island, and changes in technology, mainly electricity, changed the development at the resort. Thousands of electric lights lit the streets, rides and saloons. The lights gave North Beach a dazzling appearance, but also attracted a different kind of crowd. While families and Sunday school groups still came during the afternoon, the nightlife drew crowds of single working-class men and women who went to the saloons and dance halls.

Finally, those establishments concentrated on attracting these new customers rather than appealing to families bringing their picnic baskets. Fights and other disturbances happened more often, and the women alone in the saloon raised fears of prostitution. Although police officers had always been present at the resort, the new threats of crime for the customers created a need for more protection.

The period of North Beach's great popularity came to a fast end when Prohibition became law in 1919. Most of the legal businesses closed their doors. The people stayed away because there was no liquor to be had at a "dry" amusement park . . .

QUESTIONS: IDEAS TO THINK ABOUT

1. *Why does the ad read "North Beach, L.I." when it was really part of New York City?*
2. *What entitles a resort to be called a "family" resort? What would be a "singles" resort?*
3. *What's the impact of new technology (i.e. electricity) on recreation?*
4. *What kind of an event today would compare to a "trolley party"?*

NORTH BEACH, 1892

This view of North Beach some years after it opened shows the number of rides and attractions at the resort. Under the dome stood Kremer's carrousel, featuring wooden animals carved by Kremer himself. The carrousel, ferris wheel, and scenic railway (an early kind of roller coaster) was built by individuals on land leased from the Beach Company. Separate steam engines powered each of the rides.

NORTH BEACH, 1892.

The pony ride in the foreground moved after 1900 to the far end of the beach, and another dance hall was built on the site. As the resort's clientele changed from families to young men and women, operators discovered that there was more profit in selling beer than offering pony rides.

Across the street from the carrousel stood Corey's Hotel, featuring George Ehret's Extra Beer. Not really a resort hotel, Corey's establishment was simply a saloon and dance hall.

QUESTIONS: IDEAS TO THINK ABOUT

1. *Why would you take a picture of an empty amusement park?*
2. *Of what does the architecture remind you?*
3. *What would be the equivalent of North Beach today? How has the design of amusement parks changed (compare Disney World, Great Adventure)?*

GEO. W. KREMER'S SILVER SPRING CARROUSEL, ca. 1900. This advertisement from the "North Beach Amusement Guide" points out the kind of people Kremer hoped would ride his carrousel. Most of the rides cost a nickel (which was also the cost of the trolley and the new elevated lines), but Kremer offered "special rates" to Sunday Schools and Trolley Parties. We wonder about the "special attention" promised to "ladies" and children. Whether or not the "special attention" was real, Kremer was making a pitch in his ad for the family trade. The man standing in the center of the picture is probably Geo. W. Kremer.

"SHOOTING-THE-CHUTES" AND OTHER RIDES

The postcards and advertisements here show several of the rides built at North Beach before 1900, including the carrousels (built and carved by Kremer himself), the circle swing, the scenic railway, and, most popular of all, the Shoot - the - Chutes. The Chutes were part of a separate amusement area within the Beach known as Gala Park—notice the GP in the star on the right side of the postcard. One man whose family had a concession at North Beach from 1902 to 1916 recalls what the ride was like:

> ... a large metal ramp, it went up maybe sixty, eighty feet ... you got in a boat turned up in the front and square in the back ... this thing would go down this hill of water, pieces of wood on it would make it look foamy ... water'd splash all over you ... into an artificial lake. I was on it many times. The man who sold the tickets knew me ... On top of this at night they had a big searchlight ... quite an attraction. With all the property they had in the back, it discouraged lovers and muggings, cause this big white light went all over the property ... it was helping to police the place.

QUESTIONS: IDEAS TO THINK ABOUT

1. *Contrast the images of the pictures of the carrousel and the Shoot - the - Chutes and the description in the quotation with reference to muggings and police protection.*
2. *What does the description tell you about the concern for personal safety? How does this fit with your ideas about "the good old days"?*

CONDITIONS AT NORTH BEACH

In 1886, shortly after the resort opened, a reporter for the *Long Island Star* wrote his impressions of the crowd at Bowery Bay:

> Family parties are frequent. Of the lovers going there with sweet communion with nature the majority are thrifty looking and quiet, apparently in the stage of courtship where the man is thinking seriously about the cost of furnishing the house and the prospective bride gives him good advice about the management of his finances. The rollicking young man with his "best girl" seeking frolic only, does not seem to have discovered Bowery Bay as yet.

Remember that ideal passage as you read the following article from the *Flushing Times* of May 27, 1905, describing the immorality of the thrill-seekers at North Beach. At the end of the piece the writer attacks George Ehret, "the product of whose beer factory is sold at just about every place in town." Beer drinking was blamed for the "pandemonium" that occurred each night and the "moral ruin" of "girls of respectable parents ... caught in the whirl of excitement and crowds and music."

Articles like this blaming alcohol for the social problems helped the temperance movement to grow. The movement became a law when Prohibition was enacted in 1919. Outlawing the sale of liquor was a way of controlling behavior that not everyone agreed with. The young men and women who filled the saloons, the "toothsome crowd" described in the article, probably saw such articles as attacks on their way of life. The tone of this article is similar to the 1885 article condemning the immoral goings-on along the shore of Bowery Bay before the resort was built. In both instances the articles attack a common pattern of working-class leisure.

CONDITIONS AT NORTH BEACH

Flushing Times, May 27, 1905

Every Sunday the place swarms with the lowest class of East Side toughs and girls young in years and old in vice. Nevertheless we see girls of respectable parents going there perhaps out of curiousity, and therafter, caught by the whirl of excitement and crowds and music, each succeeding visit is but one step nearer moral ruin. The metropolitan policemen who are detailed to North Beach each Sunday are unanimous in say that it is the worst place and the most toothsome crowd they ever encountered. These men who have seen every kind of vice or criminal which the great city could produce are competent to know and judge what they are talking about. The Bowery or Coney Island in their palmy days were not the equal of North Beach for the evil influence it exerts on growing boys and girls. Any boy or girl who frequents North Beach is lost so far as her influence for good in this world is concerned. To one walking along the main thoroughfare the real conditions in the various dance halls and dives are not apparent and in fact the devil to pay does not get into full swing until dark. From that time to midnight pandemonium reigns. No attempt whatever is made to observe the Sunday liquor laws and persons of all ages are served with the vilest of compounds. Of course the real harm originates in these dance halls and it would open the eyes of some of our people if they could spend a few hours at this pest hole right at their doors. Something should be done to shut this abomination for all time. Mr. George Ehret, the product of whose beer factory is sold at just about every place in town, is largely interested in the money end of this place, but I am sure he knows nothing of the conditions existing there other than how many kegs of beer are sold every week.

BOARDWALK AND WATERFRONT, NORTH BEACH, c. 1900.

QUESTIONS: IDEAS TO THINK ABOUT

1. *What happened in the thirteen years between the 1892 picture and the 1905 article to change the public perception of North Beach?*
2. *Is the article talking about changing morality or did the groups of people who went to North Beach change in terms of class, or ethnic background?*
3. *Has the function and kind of newspaper reporting changed? Would you see an article like this today?*

FIRE WORKS AT NORTH BEACH

The weekly fireworks display was one of the most popular events at North Beach. The spotlight atop the shoot-the-chutes and the strings of electric lights on the rides reflected in the water add to the scene. A rowboat carried the fireworks to a special pier in the bay, where they would cause the least damage. One man remembers the displays he saw as a boy: "... all kinds of pinwheels ... and there'd be one masterpiece, maybe a giant glass of beer with foam coming off it, Fourth of July an American flag ... once in a while they'd have a misfortune, blow something up."

A newspaper report from 1886 mentioned that the "set pieces" drew applause from the public, especially the one that blazed "Welcome W. Steinway & Sons." The account mentions that "a slight accident occurred during the exhibition." A boy named Charles Langyard "persisted, as Mr. Newton, the manager of the beach, states, in approaching too near the fireworks and one of the rockets struck him in the right leg. The wound was not serious and after being attended by Dr. Trask, the boy was removed to his home."

This account implies that the Beach Company has the public's safety and pleasure in mind at all times, and that the boy's injury was his own fault, and not due to the Company's negligence. In any event, a doctor was ready to attend to the emergency. Accidents like this, however, probably convinced them to move the fireworks onto the pier in the bay, reducing the danger to the people and the buildings.

FIRE WORKS AT NORTH BEACH, 1886.

QUESTIONS: IDEAS TO THINK ABOUT

1. *What are the issues to think about in relation to improvements in technology and personal safety?*
2. *Technology usually first appears as amusement. Can you think of some new development in your lifetime that first came into your life as an amusement and was later taken more seriously, like the automobile?*
3. *Why do we take the symbols of war ("rocket's red glare, bombs bursting in air") and turn them into content for public displays?*

FIRE

Wood structures, gas lights, primitive electrical wiring, busy kitchens, carelessly tossed cigars and matches made fire a real danger in North Beach. In its short history the resort had many fires which destroyed rides, dance halls, and bath houses. It was fire that finally ended the era of North Beach in 1924. The first fire occurred only two months after the resort opened in 1886. *The Long Island Star* reported the incident on August 27th of that year.

> The destructive fire which took place at Bowery Bay on Thursday night will not have any material effect upon the facilities of the management to cater to the fullest extent to the wants of their patrons. Even while the ruins were still smouldering on Friday, workmen were busily engaged in building a temporary kitchen on the spot where the old one stood and at noon time the culinary department was running again almost as perfect as ever. Mr. William Steinway and Mr. Ehret visited the scene during the afternoon and partook of dinner which was admirably prepared in the kitchen then in the process of erection.

NORTH BEACH VOLUNTEER FIRE COMPANY, ca. 1910. The men are standing in front of the power station that provided electricity for North Beach.

The situation was particularly dangerous because the resort was not near any communities that could provide protection, and the fires could spread rapidly. In response to this problem the Beach Company formed its own volunteer fire company. Most of the members were concessionaires and their employees. The company was more than a fire-fighting brigade. It was a combined social and political club as well. The volunteer fire company offered the men who owned and operated North Beach the opportunity to work and relax together. The organization developed a sense of community, fraternity, and mutual respect.

POLICE PROTECTION

The safety and security of the visitors to the resort was as much of a concern of the Beach Company as their comfort and pleasure. Since Bowery Bay was outside the city limits—Queens did not become part of New York City until 1898—they hired their own police force. Indeed, all of the newspaper articles about Bowery Bay and the Beach Company's own publications stressed the efficiency and authority of Captain John Corrigan's police, who "keep a sharp look out for any objectionable intruders on the grounds . . . and if any such appear, they are quietly but decidedly told to 'quit'" (*L.I. Star*, June 11, 1886). We should ask by what criteria the police defined "objectionable intruders." It makes us wonder about the conditions at other resorts and the images in the public's mind that made such reassurances necessary.

NEW YORK CITY POLICEMEN, ca. 1915. During the week only two officers patrolled North Beach, but on weekends, when thousands of people flocked there three or four men were added. This was in addition to the resort's own men. The officers pictured here are standing on the corner where Kells' movie house stood. Bowery Bay is in the background, and the fireworks pier is visible between the two men.

The following selection is from a promotional booklet published by the Beach Company in 1889.

"The loveliness of this favored spot and the beautiful developments in which art has contributed to assist nature, would, of themselves suffice to induce tens of thousands of summer visitors to cast care aside and hasten to enjoy its charming features. Hence Bowery Bay managers, in securing absolute immunity from any possible annoyance of their patrons, have added a substantial reason to the many natural beauties of their surroundings why the public should avail themselves of the charms which court citizens from sultry homes to the cooling resorts of this heaven-favored beach; and the people of New York but to be made aware of this fact in order to obtain for Bowery Bay such a rush of patronage as will more than surprise those who cater to the public pleasure. A sense of security is the first element of civilization. Families desire innocent enjoyment, without unseemly hindrance or interruption. They will go to the places which offer the maximum of pleasure with the minimum of annoyance, and in this respect there can be no doubt that Bowery Bay Beach is beyond comparison."

From a promotional booklet published by the Bowery Bay Beach Improvement Company, 1889

NORTH BEACH POLICE STATION, ca. 1900.

QUESTIONS: IDEAS TO THINK ABOUT

1. *Do you feel that amusement has built-in risks? Can you give examples (automobile racing, etc.)?*
2. *In reading these passages, how do the references to fire and police change your ideas of the past?*
3. *Have you ever been a member of a volunteer fire or police company?*
4. *What was the function of social clubs beyond the Saturday night dances? What else did they do? (example: death benefits).*

This excerpt from Stephen Crane's short story of life on the streets of New York in the 1890s shows how quickly violence could erupt in a saloon. In the story, Jimmy fought for the honor of his sister, Maggie, who had been out the previous night with Pete, the bartender.

Precisely such scenes as this gave saloons a bad reputation—at least among the "respectable" people who did not frequent them—and strengthened the temperance movement. Saloons were blamed for many urban problems, ranging from drunkenness, crime, and violence to political corruption (the ward "boss" often used saloons as their headquarters) and prostitution, especially after the Rain's Law was passed.

MAGGIE, A GIRL OF THE STREETS
by Stephen Crane, 1983

On a corner, a glass-fronted building shed a yellow glare upon the pavements. The open mouth of a saloon called seductively to passengers to enter and annihilate sorrow or create rage.

The interior of the place was papered in olive-and-bronze tints of imitation leather. A shining bar of counterfeit massiveness extended down the side of the room. Behind it, a great mahogany-imitation sideboard reached the ceiling. Upon its shelves rested pyramids of shimmering glasses that were never disturbed. Mirrors set in the face of the sideboard multiplied them. Lemons, oranges, and paper napkins, arranged with mathematical precision, sat among the glasses. Many-hued decanters of liquor perched at regular intervals on the lower shelves. A nickel-plated cash register occupied a place in the exact center of the general effect. The elementary senses of it all seemed to be opulence and geometrical accuracy.

Across from the bar, a smaller counter held a collection of plates which swarmed frayed fragments of crackers, slices of boiled ham, disheveled bits of cheese, and pickles swimming in vinegar. An odor of grasping, begrimed hands and munching mouths pervaded all.

Pete, in a white jacket, was behind the bar, bending expectantly toward a quiet stranger. "A beeh," said the man. Pete drew a foam-topped glassful, and set it dripping upon the bar.

At this moment, the light bamboo doors at the entrance swung open and crashed against the wall. Jimmie and a companion entered. They swaggered unsteadily but belligerently toward the bar, and looked at Pete with bleared and blinking eyes.

"Gin," said Jimmie.

"Gin," said the companion.

Pete slid a bottle and two glasses along the bar. He bent his head sideways as he assiduously polished away with a napkin at the gleaming wood. He wore a look of watchfulness. Jimmie and his companion kept their eyes upon the bartender and conversed loudly in tones of contempt.

"He's a dandy masher, ain't he?" laughed Jimmie.

"Well, ain't he!" said the companion, sneering. "He's great, he is. Git on to deh mug on deh blokie. Dat's enough to make a feller turn handsprings in 'is sleep."

The quiet stranger moved himself and his glass a trifle farther away and maintained an attitude of obliviousness.

"Gee! Ain't he hot stuff?"

"Git on to his shape!"

"Hey!" cried Jimmie, in tones of command. Pete came along slowly, with a sullen dropping of the underlip.

"Well," he growled, "what's eatin' yehs?"

"Gin," said Jimmie.

"Gin," said the companion.

As Pete confronted them with the bottle and the glasses, they laughed in his face. Jimmie's companion, evidently overcome with merriment, pointed a grimy forefinger in Pete's direction. "Say, Jimmie," demanded he, "what's dat behind d' bar?"

"Looks like some chump," replied Jimmie. They laughed loudly.

Pete put down a bottle with a bang and turned a formidable face toward them. He disclosed his teeth, and his shoulders heaved restlessly. "You fellers can't guy me," he said. "Drink yer stuff an' get out an' don' make no trouble."

Instantly the laughter faded from the faces of the two men, and expressions of offended dignity immediately came. "Aw, who has said anyt'ing t' you?" cried they in the same breath.

The quiet stranger looked at the door calculatingly.

"Ah, come off," said Pete to the two men. "Don't pick me up fer no jay. Drink yer rum an' git out an' don' make no trouble."

"Aw, go ahn!" airily cried Jimmie.

"Aw, go ahn!" airly repeated his companion.

"We goes when we git ready! See?" continued Jimmie.

"Well," said Pete in a threatening voice, "don' make no trouble."

Jimmie suddenly leaned forward with his head on one side. He snarled like a wild animal. "Well, what if we does? See?" said he.

Hot blood flushed into Pete's face, and he shot a lurid glance at Jimmie. "Well, den we'll see who's d' bes' man, you or me," he said.

The quiet stranger moved modestly toward the door.

Jimmie began to swell with valor. "Don' pick me up fer no tenderfoot. When yeh tackles me, yeh tackles one of d' bes' men in d' city. See? I'm a scrapper, I am. Ain't dat right, Billie?"

"Sure, Mike," responded his companion in tones of conviction.

"Aw!" said Pete, easily. "Go fall on yerself."

The two men again began to laugh.

"What is dat talking?" cried the companion.

"Don' ast me," replied Jimmie with exaggerated contempt.

Pete made a furious gesture. "Git outa here now, an' don' make no trouble. See? Youse fellers er lookin' fer a scrap, an' it's like yeh'll fin' one if yeh keeps on shootin' off yer mout's. I know yehs! See? I kin lick better men dan yehs ever saw in yer lifes. Dat's right! See? Don' pick me up fer no stiff, er yeh might be jolted out in d' street before yeh knows where yeh is. When I comes from behind dis bar, I t'rows yehs bote inteh d' street. See?"

"Ah, go ahn!" cried the two men in chorus.

The glare of a panther came into Pete's eyes. "Dat's what I said! Unnerstan'?"

He came through a passage at the end of the bar and swelled down upon the two men. They stepped promptly forward and crowded close to him. They bristled like three roosters. They moved their heads pugnaciously and kept their shoulders braced. The nervous muscles about each mouth twitched with a forced smile of mockery.

"Well, what yer goin' t' do?" gritted Jimmie.

Pete stepped warily back, waving his hands before him to keep the men from coming too near.

"Well, what yer goin' t' do?" repeated Jimmie's ally. They kept close to him, taunting and leering. They strove to make him attempt the initial blow.

"Keep back now! Don' crowd me," said Pete ominously.

Again they chorused in contempt, "Aw, go ahn!"

In a small, tossing group, the three men edged for positions like frigates contemplating battle.

"Well, why don' yeh try t' t'row us out?" cried Jimmie and his ally with copious sneers.

The bravery of bulldogs sat upon the faces of the men. Their clenched fists moved like eager weapons. The allied two jostled the bartender's elbows, glaring at him with feverish eyes and forcing him toward the wall.

Suddenly Pete swore furiously. The flash of action gleamed from his eyes. He threw back his arm and aimed a tremendous, lightning-like blow at Jimmie's face. His foot swung a step forward and the weight of his body was behind his fist. Jimmie ducked his head, Bowery-like, with the quickness of a cat. The fierce answering blows of Jimmie and his ally crushed on Pete's bowed head.

The quiet stranger vanished.

The arms of the combatants whirled in the air like flails. The faces of the men, at first flushed to flame-colored anger, now began to fade to the pallor of warriors in the blood and heat of a battle. Their lips curled back and stretched tightly over the gums in ghoul-like grins. Through their white, gripped teeth struggled hoarse whisperings of oaths. Their eyes glittered with murderous fire.

Each head was huddled between its owner's shoulders, and arms were swinging with marvelous rapidity. Feet scraped to and fro with a loud, scratching sound upon the sanded floor. Blows left crimson blotches upon the pale skin. The curses of the first quarter-minute of the fight died away. The breaths of the fighters came wheezing from their lips and the three chests were straining and heaving. Pete at intervals gave vent to low, labored hisses, that sounded like a desire to kill. Jimmie's ally gibbered at times like a wounded maniac. Jimmie was silent, fighting with the face of a sacrificial priest. The rage of fear shone in all their eyes, and their blood-colored fists whirled.

At a critical moment, a blow from Pete's hand struck the ally, and he crashed to the floor. He wriggled instantly to his feet and, grasping the quiet stranger's beer glass from the bar, hurled it at Pete's head.

High on the wall it burst like a bomb, shivering fragments flying in all directions. Then missiles came to every man's hand. The place had heretofore appeared free of things to throw, but suddenly glasses and bottles went singing through the air. They were thrown point-blank at bobbing heads. The pyramids of shimmering glasses, that had never been disturbed, changed to cascades as heavy bottles were flung into them. Mirrors splintered to nothing.

The three frothing creatures on the floor buried themselves in a frenzy for blood. There followed in the wake of missiles and fists some unknown prayers, perhaps for death.

The quiet stranger had sprawled very pyrotechnically out on the sidewalk. A laugh ran up and down the avenue for the half of a block. "Dey've t'rowed a bloke inteh deh street."

People heard the sound of breaking glass and shuffling feet within the saloon and came running. A small group, bending down to look under the bamboo doors, and watching the fall of glass and three pairs of violent legs, changed in a moment to a crowd. A policeman came charging down the sidewalk and bounced through the doors into the saloon. The crowd bent and surged in absorbing anxiety to see.

Jimmie caught the first sight of the oncoming interruption. On his feet he had the same regard for a policeman that, when on his truck, he had for a fire engine. He howled and ran for the side door.

The officer made a terrific advance, club in hand. One comprehensive sweep of the long night stick threw the ally to the floor and forced Pete to a corner. With his disengaged hand he made a furious effort at Jimmie's coattails. Then he regained his balance and paused. "Well, well, you are a pair of pictures. What have ye been up to?"

Jimmie, with his face drenched in blood, escaped up a side street, pursued a short distance by some of the more law-loving or excited individuals of the crowd.

Later, from a safe, dark corner, he saw the policeman, the ally, and the bartender emerge from the saloon. Pete locked the doors and then followed up the avenue in the rear of the crowd-encompassed policeman and his charge.

At first Jimmie, with his heart throbbing at battle heat, started to go desperately to the rescue of his friend, but he halted. "Ah, what's d' use?" he demanded of himself.

PROHIBITION AND THE DECLINE OF NORTH BEACH

When Prohibition became law in 1919, the effect on North Beach was devastating. The breweries that owned the saloons and many of the other legitimate entrepreneurs walked away from North Beach, and the resort became almost a ghost town. The rides and amusements and hotels that could not serve beer were not enough to draw the crowds. After a year or two, bootleggers moved into a few of the buildings and soon they were making more money selling watery beer and flavored grain alcohol than the legitimate operators ever did.

The law was in fact a direct attack on the leisure activities that were so very popular at North Beach. Beer gardens, dance halls, and saloons were seen not as the normal recreational centers of the largely immigrant working-class, but as places that corrupted the morals of those who frequented them. Despite this negative aspect of the law, it did have a positive effect in reducing public intoxication and alcoholism and the resulting family and social problems. Prohibition did not mean that Americans stopped drinking, only that their drinking habits changed.

PRE-PROHIBITION ANTI-SALOON LEAGUE POSTER,

Photograph courtesy of American History Slide Collection.

ANTI-SALOON LEAGUE POSTER, ca. 1910

With ratification of the Eighteenth Amendment to the Constitution in 1919, Prohibition became the law of the land. The rise of the Temperance movement and the drive against working-class saloons by social and political reformers characterized the preceding decades. The efforts of such organizations as the Women's Christian Temperance Union and the Anti-Saloon League culminated in the "Noble Experiment," as President Herbert Hoover would later call Prohibition.

This Anti-Saloon League poster portrays the saloon's threat to family life, represented by the innocent little girl clutching her doll. In the text, the poster links wholesome family values with Christian ideals and patriotic sentiments.

QUESTIONS: IDEAS TO THINK ABOUT

1. *Why does an anti-saloon poster use the image of a man's hand threatening a small girl holding a doll?*
2. *What is the poster really trying to attack? Are they trying to destroy selling liquor or neighborhood bars?*
3. *Is this document really a religious argument ("Our religion demands")? Are we looking at the native Protestant values attacking the immigrant population?*

BIG BLONDE
by Dorothy Parker

In the passage below, from a short story about alcoholism, Dorothy Parker gives us a glimpse into the speakeasy which flourished during Prohibition. Ed is taking Mrs. Morse, his mistress, to a speakeasy named Jimmy's. Notice, in Parker's descriptions of the women who frequented this bar, the underlying mood of loneliness and despair—and the ever ready solution for the sadness, a "little drink."

"BIG BLONDE" BY DOROTHY PARKER

Ed brought her to Jimmy's. He was proud, with the pride of the transient who would be mistaken for a native, in his knowledge of small, recent restaurants occupying the lower floors of shabby brownstone houses; places where, upon mentioning the name of an habitué friend, might be obtained strange whisky and fresh gin in many of their ramifications. Jimmy's place was the favorite of his acquaintances.

There, through Ed, Mrs. Morse met many men and women, formed quick friendships. The men often took her out when Ed was in Utica. He was proud of her popularity.

She fell into the habit of going to Jimmy's alone when she had no engagement. She was certain to meet some people she knew, and join them. It was a club for her friends, both men and women.

The women at Jimmy's looked remarkably alike, and this was curious, for, through feuds, removals, and opportunities of more profitable contacts, the personnel of the group changed constantly. Yet always the newcomers resembled those whom they replaced. They were all big women and stout, broad of shoulder and abundantly breasted, with faces thickly clothed in soft, high-colored flesh. They laughed loud and often, showing opaque and lusterless teeth like squares of crockery. There was about them the health of the big, yet a slight, unwholesome suggestion of stubborn preservation. They might have been thirty-six or forty-five or anywhere between.

They composed their titles of their own first names with their husbands' surnames—Mrs. Florence Miller, Mrs. Vera Riley, Mrs. Lilian Block. This gave at the same time the solidity of marriage and the glamour of freedom. Yet only one or two were actually divorced. Most of them never referred to their dimmed spouses; some, a shorter time separated, described them in terms of great biological interest. Several were mothers, each of an only child—a boy at school somewhere, or a girl being cared for by a grandmother. Often, well on toward morning, there would be displays of kodak portraits and of tears.

They were comfortable women, cordial and friendly and irrepressibly matronly. Theirs was the quality of ease. Become fatalistic, especially about money matters, they unworried. Whenever their funds dropped alarmingly, a new donor appeared; this had always happened. The aim of each was to have one man, permanently, to pay all her bills, in return for which she would have immediately given up other admirers and probably would have become exceedingly fond of him; for the affections of all of them were, by now, unexacting, tranquil, and easily arranged. This end, however, grew increasingly difficult yearly. Mrs. Morse was regarded as fortunate.

Ed had a good year, increased her allowance and gave her a sealskin coat. But she had to be careful of her moods with him. He insisted upon gaiety. He would not listen to admissions of aches or weariness.

"Hey, listen," he would say, "I got worries of my own, and plenty. Nobody wants to hear other people's troubles, sweetie. What you got to do, you got to be a sport and forget it. See? Well, slip us a little smile, then. That's my girl."

She never had enough interest to quarrel with him as she had with Herbie, but she wanted the privilege of occasional admitted sadness. It was strange. The other women she saw did not have to fight their moods. There was Mrs. Florence Miller who got regular crying jags, and the men sought only to cheer and comfort her. The others spent whole evenings in grieved recitals of worries and ills; their escorts paid them deep sympathy. But she was instantly undesirable when she was low in spirits. Once, at Jimmy's, when she could not make herself lively, Ed had walked out and left her.

"Why the hell don't you stay home and not go spoiling everybody's evening?" he had roared.

Even her slightest acquaintances seemed irritated if she were not conspicuously light-hearted.

"What's the matter with you, anyway?" they would say. "Be your age, why don't you? Have a little drink and snap out of it."

THE AIRPORT, 1934-1939

In spite of the decline after Prohibition, parts of North Beach remained until the middle of the 1930's, when construction began on LaGuardia Airport. The swimming pool was still open; people continued to use the beach; the trolley still stopped there and there was ferry service in the summer. There is no question, however, that North Beach was only a shell, and planners and developers saw that the area could be put to better, more efficient use.

During the 1920's North Beach began its association with air travel. The site first attracted attention as a seaplane base and within a few years a regular airfield was built, though not on a scale to handle large commercial aircraft. At that time New York City's only municipal airport was Floyd Bennet Field at the end of Flatbush Avenue in Jamaica Bay, not very convenient to downtown. When Mayor LaGuardia decided to build a new and completely modern airport, his attention was naturally drawn to North Beach. The advantages of this site were: an airfield already in operation; suitability for transatlantic seaplanes. The new Grand Central Parkway and Triborough Bridge brought midtown Manhattan close to the airport.

Today, people still go to the shore of Bowery Bay in pursuit of leisure. Now they depart from there for vacations to such places as Florida, the Caribbean. It requires no great leap of the imagination to see Disneyworld in Orlando, Florida, as a direct descendant of North Beach in Queens, Long Island. Just as the streetcar companies offered special excursions to North Beach, including for one price transportation, food, beer and entertainment, so today the airlines offer complete vacation packages to Orlando.

SEAPLANES ON BOWERY BAY

One of the earliest events in the development of aviation at North Beach was the landing of the German DO-X seaplane in the spring of 1929. The six-engine aircraft flew non-stop across the Atlantic and landed on Bowery Bay because there was already a small seaplane base there. The landing had been built by Carl Deutschmann, son of the man who owned Silver Spring.

His business (except for the swimming pool which lasted until the mid-1930s) had been ruined by Prohibition, and he introduced seaplane rides in an effort to attract customers.

The DO-X immediately became a tourist attraction. Thousands of New Yorkers made the trip to North Beach to view the famous aircraft, and many of them bought tickets for a tour of the plane.

Seaplanes like the DO-X were thought to be the future of transatlantic air travel because no airports on land were then big enough to handle large commercial airliners. They also could take advantage of the unrestricted space of large bodies of water and could adjust their take-offs and landings to wind conditions. With improvements in flight technology, however, the seaplane's historical moment passed.

When the city decided to build an international airport at North Beach they made plans for seaplane service. The newspaper article below describes the departure of the first Pan-American Clipper from the Marine Air Terminal, an event witnessed by a crowd of 150,000.

DORNIER DO-X SEAPLANE ON BOWERY BAY, 1929.

QUESTIONS: IDEAS TO THINK ABOUT

1. *In what ways can we see air travel as "playing with the future"?*
2. *How and why do new inventions, like the airplane, stimulate fantasy?*
3. *Is there a connection between inventors and artists?*

GLENN CURTIS AIRPORT, 1934

With the increasing popularity and importance of flight, North Beach became a center of aviation in Queens. In addition to the seaplane base, the 1920's also saw the construction of Glenn Curtis Airport, a privately owned and operated field. To make way for the runway and hangers, many of the remaining structures from the amusement park were torn down, leaving only the swimming pool.

This airfield was not sufficient to handle large-scale commercial traffic, however. The following editorial from the *New York Evening Journal* of February 26, 1937, decries the city's lack of airport facilities while boosting North Beach as the most desirable location to build one. The editorial makes an interesting connection with the 1939 World's Fair in nearby Flushing Meadows, optimistically predicting that "thousands of exposition visitors will arrive by air."

N. Y. Journal
FRIDAY, FEBRUARY 26, 1937
Published Daily, Except Sunday by New York Evening Journal, Inc., at 220 South Street, New York City.

Airportless New York

North Beach Ideal Site for Trans-Atlantic Mail, Passenger Terminus

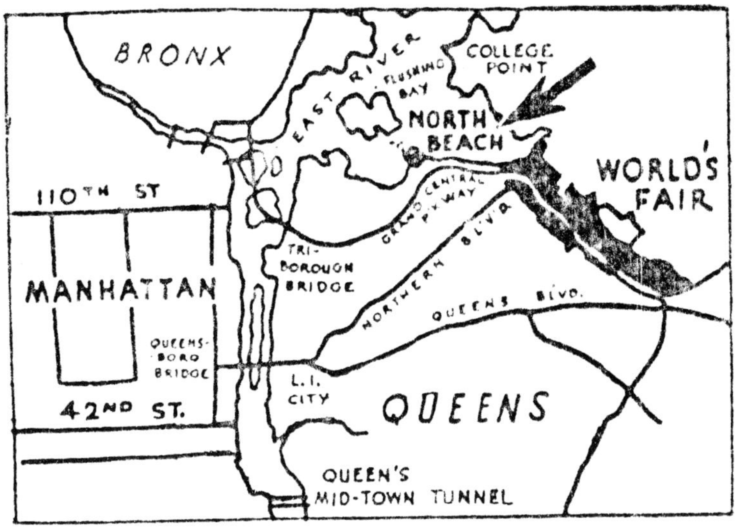

With the probable selection of North Beach Airport as the western terminal of regular trans-Atlantic air service, New York at last seems to be coming into its rightful own.

This city of yours, the largest in America, now has no airmail field.

It should have.

It should have had one long ago.

And if the city and Federal Government combine to develop North Beach it will have one of the greatest, if not **THE** greatest, mail and passenger airport in the world.

With Pan-American Airways and the Imperial Airways of Great Britain as the operating companies, transatlantic service is to begin on November 1.

Four round trips a week are scheduled.

The Federal Government has set aside $3,500,000, the city will be required to appropriate $1,500,000; other millions will eventually be needed to enlarge, equip and modernize the field.

Not the least feature of the splendid air future now open to New York is the fact that the airport is near the site of the 1939 World's Fair in Flushing Meadows.

Thousands of exposition visitors will arrive by air. At North Beach they will be able to step from a plane almost at the gate of the Fair, on a field seemingly designed by nature itself for an airport.

It is difficult to conceive of a more nearly ideal site for both land and sea planes, such as are now using it.

When completed it will have 3,500-foot runways and practically unlimited take-off space over the waters of the widening East River and Long Island Sound.

There are no tall buildings near it, the waters themselves will serve as guides, and with the completion of the Manhattan-Queens Tunnel, access to the heart of town will be but a matter of minutes.

A greater contrast with our present passenger and airmail service—or lack of service—cannot be imagined. The city must have this improvement. It means millions to the people of New York.

Airport a Magnet

New York played "let's-go-to-the-new-airport" game to the tune of 150,000 yesterday, aided and abetted by a balmy touch to the weather. Police handled crowds at LaGuardia Field at North Beach that came to see the sky giants reclining and soaring. Here spectators line a ramp to watch the goings-on.

— *Story on page 10; other picture on page 1.*

The new airport opened on December 10, 1939, and the public soon discovered that going to the new airport was itself a different way to spend a Sunday afternoon. They were drawn by the sense of the "new," and it was exciting to watch the airliners take off and land, even though few of the people in the crowd could afford to fly at that time; in fact, many of them were still experiencing the effects of the worst depression in American history. How did they react in that context to the new airport? We wonder if this look at the future gave them a sense of optimism.

QUESTIONS: IDEAS TO THINK ABOUT

1. *How does having an airport nearby affect a community?*
2. *Why does a leisure site, like North Beach, become an airport?*
3. *North Beach Airport, later LaGuardia, maintained the function of that land, but it served a different group of people. Who was the new community that it served?*
4. *How can you explain the expenditure of Federal funds in the development of an airport in 1937 at a time of great depression?*

RABBIT IS RICH

Free! Macadam falls away beneath the wheels, a tawny old fort can be glimpsed as they lift off the runway beneath the rounded riveted edge of one great wing, the gas tanks of South Philadelphia are reduced to a set of white checkers. The wheels thump, retracted, and cruel photons glitter on the aluminum motionless beside the window. The swift ascent of the plane makes their blood weighty; Janice's hand sweats in his. She had wanted him to have the window seat, so she wouldn't have to look. There is marsh below, withered and veined with saltwater. Harry marvels at the industrial buildings beyond the Delaware: flat gravel roofs vast as parking lots and parking lots all inlaid with glittering automobile roofs like bathroom floors tiled with jewels. And in junkyards of cars the effect is almost as brilliant. The NO SMOKING sign goes off. Behind the Angstroms the voices of the Murketts and the Harrisons begin to chatter. They all had a drink at an airport bar, though the hour was eleven in the morning. Harry has flown before, but to Texas with the Army and dealers' conferences in Cleveland and Albany: never aloft on vacation like this, due east into the sun. How quickly, how silently, the 747 eats up the toy miles below! Sun glare travels with them across lakes as momentarily as across a mirror. The winter has been eerily mild thus far, to spite the Ayatollah; on golf courses the greens show as living discs and ovals amid the white beans of the traps and on the fairways he can spot moving specks, men playing. Composition tennis courts are dominoes from this height, drive-in movies have the shape of a fan, baseball diamonds seem a species of tattered money. Cars move very slowly and with an odd perfection, as if the roads hold tracks. The houses of the Camden area scatter, relenting to disclose a plowed field or an estate with its prickly mansion and eye of a swimming pool tucked in the midst of woods; and then within another minute, still climbing, Harry is above the black-red carpet of the Jersey Pines, scored with yellow roads and patches of scraping but much of it still unmarred, veins of paler unleafed trees following the slope of land and flow of water among the darker evergreens, the tints of competition on earth made clear to the eye so hugely lifted. Janice lets got of his hand and gives signs of having swallowed her terror.

"What do you see?" she asks.

"The Shore."

It is true, in another silent stride the engines had inched them to the edge of the ocean of trees and placed underneath them a sandy strip, separated from the mainland by a band of flashing water and filled to a precarious fullness with linear summer cities, etched there by builders who could not see, as Harry can, how easily the great shining shoulder of the ocean could shrug and immerse and erase all traces of men. Where the sea impinges on the white sand a frill of surf slowly waves, a lacy snake pinned in place. Then this flight heads over the Atlantic at an altitude from which no whitecaps can be detected in the bluish hemisphere below, and immensity becomes nothingness. The plane, its earnest droning without and its party mutter and tinkle within, becomes all of the world there is.

John Updike

TRANSPORTATION

INVITATION TO SEEING IN PERSPECTIVE

This chapter is a collection of documents, literature, and impressions about transportation and about people. Two themes run through this chapter. One is the historical process, the way one form of transportation replaces another, and the impact of that change on people's lives. The second is a clash of perspectives: a question of good and bad—the competing values that are part of change over time. Of course, how people perceive change depends on their point of view.

Riding a trolley or driving a car creates different impressions depending on one's perspective. A ferry offers a leisurely boat ride in the fresh salt air, or in winter a slow and icy way to commute over the river. A bridge is a triumph of man's vision over the forces of nature, or the beginning of problems in rapid urban development. The concerns of the businessman and the stirrings of the artist, the experience of the worker and the words of the poet—all are part of this perspective.

Different forms of transportation operate in the same place and at the same time, while some gain in popularity and ridership others decline. The Queensboro Bridge and the Triboro Bridge for example, made the East River ferries obsolete. Up to World War II Queens was primarily a streetcar suburb; the bus replaced the trolley and Queens developed around surface transit.

Transportation in the past had its own set of problems; as well as being often crowded and inconvenient, it was many times open to the effects of the weather. Why should we assume it was better then than now? Today we may romanticize the horse and buggy, but we often neglect to recognize the dirt, maintenance, and health problems associated with it. When the 92nd Street ferry went out of business in 1918 due to competition from the Queensboro Bridge, the people of Astoria pressed the city to take over its operation. In 1978 residents of Howard Beach, protesting the proposed landing of the Concorde SST at Kennedy Airport, created a massive traffic jam at the airport. Both groups attempted to control the effects of transportation changes on their lives. Because each new form of travel affects people in their communities, by thinking about these changes we can learn about the role of transportation in people's lives at home and the workplace.

THE HORSE AND BUGGY

Sparsely settled in comparison with heavily urbanized Manhattan, Queens remained a rural county of farms and small towns until the first decade of the twentieth century. The towns were connected by trolley lines. The road system was relatively rural: a maze of dirt roads that grew from old Indian trails, eighteenth century farm roads, and nineteenth century toll roads—long, straight roads built and maintained by private companies.

For wealthier people, the horse-drawn wagon in its functional simplicity or a buggy in regal splendor—"the surrey with the fringe on top"—was the only way to travel over those roads, to cover distances too far to walk or not served by a private horsecar or railroad line.

We will now examine the different ways Queens' residents used horse-drawn vehicles for work and family travel.

BAYSIDE 1906. A man and woman pose proudly in front of their general store in 1906, when small towns and farms made up most of Queens. Behind the wooden building is farmland, and looking carefully, one sees a bicycle and a wagon wheel.

QUESTIONS: IDEAS TO THINK ABOUT

1. *Looking at this photograph, describe what life was like in the early 1900's. What details in the picture support your ideas?*
2. *How does having a store in the same building as a home affect one's lifestyle? Have you ever lived above a store?*
3. *How would this street look today?*

THE GEIPEL FAMILY

Nothing better illustrates the change from the nineteenth century experience to the twentieth than these photographs of the Geipel family. In 1900, a carriage was a mark of status. Fifteen years later a touring car replaced the horse and buggy as a status symbol. Even today, both adolescents and people moving up the economic ladder see a car as a mark of success.

WILHEMINA GEIPEL, HER HOUSE AND CARRIAGE, 1900. Wilhemina Geipel, a midwife, sits in her buggy with her grandchildren. She travelled by buggy to homes in the neighborhood where she delivered babies. A carriage was both a necessity for her profession and evidence of her standing and wealth in the Astoria Community.

Wilhelmine Geipel worked as a midwife, serving the German community in Astoria. A few days after she had been to a home and delivered a baby, Mrs. Geipel often sent one of her small grandchildren to collect her $10 fee. Since the family no longer needed the midwife's services, it was not uncommon for them to refuse to pay, and some even threw buckets of water on the children.

THE GEIPEL FAMILY, 1915. This automobile replaced the horse and buggy, and the barn is now a garage.

The carriage was both a necessity for Mrs. Geipel's profession and evidence of her standing and wealth. Her role in the community would have changed considerably had she made her rounds by car instead of by horse and buggy. Riding in her buggy, she was more intimately a part of her neighborhood; she could stop and greet people she recognized on the street. In the faster car she would probably miss that opportunity for personal contact with her neighbors, but at the same time she could serve more people in a shorter time and with less effort.

JOHN CHINKEL AND RENKEN'S DAIRY — BLACKSMITH TO AUTO MECHANIC

Transporting fresh products like milk, bread and vegetables to people provided work for drivers, stable hands, carpenters, blacksmiths and others. Livery stables employed unskilled workers in Queens, and many immigrants found jobs in those places.

John Chinkel was such a person. When he arrived in New York from Austria in 1925 he immediately went to the German community of Glendale where he found a room. Eventually he married his landlady's daughter. It was common for new settlers to use the immigrant network to find jobs, living space, and a social life. Armed with his landlady's advice to follow the dairy wagon, the next day he set out to find work. He stood in the busy courtyard of Renken's Dairy on Myrtle Avenue and "old man" Renken came up to him and asked, in German, whether he was looking for work. John said yes, and Renken handed him a shovel to clean the manure out of the stables. Within a short time, Chinkel began to work on repairing the wagons and soon worked his way up to chief of the maintenance department.

In about 1935 Renken bought a pair of diesel engines from Germany and installed them in his plant. They produced enough power for all of the refrigeration units and lights in his dairy. He also replaced the horse drawn wagons with a fleet of new electric trucks. Chinkel's responsibilities shifted, then, from wagon maintenance to truck maintenance and each night he had to charge the trucks' batteries with electricity produced by diesels.

John Chinkel and the other workers at Renken's Dairy had to adapt their skills to new technology. It seems odd to us today that a man who had supervised work on wooden wagons should be put in charge of such complex machinery as diesel engines and electric trucks, instead of an "expert" being brought in. During this period of transition from the old ways to the new, workers' skills were not yet obsolete and were the base on which new skills were built.

QUESTIONS: IDEAS TO THINK ABOUT

1. *Why did the dairy switch from horse-drawn wagons to trucks? How did this change affect the workers?*
2. *In addition to milk, what other products were delivered to customers' homes? Why do we no longer see that today?*
3. *Why did the dairy buy electric rather than gasoline powered trucks? What would be advantages of electric-powered vehicles for our current transportation problems?*

DELIVERY WAGON, 1925.

Transporting fresh products like milk, bread and vegetables to people provided work for drivers, stable hands, carpenters, blacksmiths and others.

ELECTRIC TRUCKS, 1935. When Renken's Dairy replaced the horse drawn delivery wagons with electric trucks, the company could serve customers within a wider area. Since they made this change in the 1930's, when the country was in the midst of the Great Depression, they may have seen this change as a way to increase business and decrease labor costs. As increasing numbers of individuals and businesses changed from horsedrawn wagons to trucks, stables went out of business and were quickly replaced by the repair shop—usually in the same building.

THE FERRY

Before bridges spanned the rivers and bays of New York there were hundreds of ferries to carry people across. While bridges connect a road or highway on one side with one on the other, ferries often connected the communities on either side.

For passengers, ferries provide a different kind of experience than bridges and emphasize different values. Ferries offer a feeling for the open air and for flowing water, for the power of the river. For most passengers, however, the ferry meant nothing more than a way to cross from Queens to Manhattan, from home to work and back again. It was an inexpensive but often slow, crowded, inconvenient and even uncomfortable way to reach the other side. The ride, pleasant and refreshing in summer, was freezing and icy in winter.

"THE STEINWAY" UNLOADING PASSENGERS IN ASTORIA, 1922. The Astoria 92nd Ferry was a link between the German communities in Yorkville and Astoria. The fare was 3¢, and workers took the ferry to get to their jobs with the Steinway Piano Company in Queens or George Ehret's Brewery (later Jocob Ruppert's) in Manhattan.

Documents like the ferry ticket shown here, photographs, newspapers, and artifacts are the historian's tools. An historian examining the more recent past uses oral history as well, the memories of people who lived through that time. Though not always historically accurate and sometimes offering a too nostalgic look at "the good old days," oral histories do communicate a personal sense of the past.

One man who grew up in Long Island City remembers a particular ride he took on the 92nd Street ferry when he was a small boy. It seems that his father's cousin in the Bronx was remodeling his home, and offered him the old bathtub. So early one morning the boy's father and uncle got a horse and a wagon for the trip across the river and up Second Avenue to the Bronx. Along the way, they stopped in numerous saloons to quench their thirst with cold beer. When they finally reached the Bronx, they loaded the bathtub onto the wagon and headed back down to the ferry landing, again pausing in most of the saloons along the way. When they reached the ferry terminal at 92nd Street they realized that they had only enough money to get themselves and their wagon across but not for the little 5 year old boy, so they hid him under the bathtub and the gatekeeper was none the wiser.

The 92nd St. ferry actually ceased operations twice, as the following newspaper articles explain. In 1918, nine years after the Queensboro Bridge opened and two years after the start of subway service to Astoria, the private company that ran the ferry stopped all service. There was such a public outcry, from both sides of the river, that the city took over the line and ran it, unprofitably, for 16 years until the final voyage of August 1, 1936.

FERRY DENIED AID, STOPS ALL SERVICE

Company Operating Between Astoria and Manhattan Claims Loss Two Years.

The New York and East River Ferry Company, operating between 92d st., Manhattan, and Fulton ave., Long Island City, permanently discontinued all service at 9 o'clock last night. Notice to that effect was posted yesterday by official of the company on the ferry houses in Astoria and Manhattan. Up to that time the company operated three ferryboats.

Failure of the city officials to take action upon a request from the company that the city either buy the company's franchise or provide a yearly subsidy that would enable the company to continue operation without a loss resulted in the suspension of service. This request was made of the city several months ago and there have been a number of conferences. Property owners and business men both in Astoria and the Yorkville section of Manhattan have stated to the city officials that the closing of this ferry would be a big disadvantage to them. As the result of an investigation conducted by the officials of the Dock Department a recommendation was made to the Board of Estimate that some action be taken by the city to keep the ferry running. It was very convenient for hundreds of workers from the upper East Side who found employment in Queens factories.

Started in 1864, the ferry has been in continuous operation ever since. The traffic was very heavy upon it until the opening of the Queensboro bridge in 1908. In former years many Long Island farmers used the ferry on their way to public markets in Manhattan. With the disappearance of the farms and market gardens this traffic has all but disapeared. At one time the company was very prosperous. According to the statement of officials of the company operation has been continued at a loss for several years.

THE LAST VOYAGE OF THE ASTORIAL FERRY, 1936

THE LAST VOYAGE OF THE ASTORIAL FERRY, 1936.

52 Old Friends Bid Good By to Astoria Ferry

Sentimental Veterans of Crossing From 92d St. Make the Last Voyage

Boat Then Goes to Rest

Thus Ends Service Routed by Bridge After 16 Years

(Reprinted from yesterday's late editions)

As sentimental voyagers stood in the Astoria ferryhouse and clanged the old fog bell in farewell the ferry-boat Rockaway sailed down the moon-lit East River early Saturday to the layoff base of the Department of Plant and Structures, at Atlantic Avenue, Brooklyn, bringing to a close the six-teen-year service of the Astoria-Ninety-second Street ferry line.

The determined, last-ditch fight of Frederick J. H. Kracke, Commissioner of Plant and Structure, had gone for naught. Robert Moses, executive officer of the Triborough Bridge Authority, whose workmen had been chased out of the Manhattan ferryhouse last week by the police when they started to destroy the slips to make way for the southerly bridge approaches, had won out at last. The farewell voyage of the Rockaway was, in fact, only the tangible symbol of Mr. Moses's triumph, for on Wednesday the Sinking Fund Commission had voted to discontinue the ferry service Saturday.

Fifty-two on Final Crossing

The final round trip from Astoria started at 11.45 p. m. (daylight saving time) with fifty-two veterans of the crossing taking the cruise as a sentimental pilgrimage. Mrs. Hattie Arnez, who lives at 1-01 Astoria Boulevard, Astoria, played the accordion and they all sang.

The Rockaway started back from Manhattan at 12:01 a. m., with the pilgrims remaining on the boat, enabling them to make the round-trip for only a 5-cent fare. They sang "East Side, West Side," and then they sang "In the Sweet Bye and Bye." Then, with an eye on future tortuous passages by way of the Queensborough Bridge, they sang, with just concern, "Where Do We Go From Here?" At 12:15 a. m., as the Rockaway bumped into the Astoria slip for the last call, they turned to "Auld Lang Syne."

The fifty-two pilgrims disembarked and in their place came ten men and two women, who boarded the boat for the reason that they simply wanted to get to Manhattan. As Captain Levi Brodnead, the skipper, put out for the west for the last time he sounded three long and dolorous blasts on the Rockaway's whistle.

Ferryhouse Dog Remains

In the Astoria ferryhouse remained Red, an airedale that wandered in as a pup ten years ago and lived there on terms of cordiality with passengers, captains, engineers, bridgemen and deckhands. The dog was left in charge of the night watchman, who said he would care for it and take it home, although it didn't seem to be much use, for Red, with the habit of a decade strong on it, was certain to wander back to the ferryhouse no matter where it was taken.

The Rockaway, after being overhauled, will be transferred to some other ferry service of the Department of Plant and Structures. New jobs must be found for three captains, three engineers and fifteen bridgemen and deckhands, the oldest of whom, in point of service, is Daniel J. McCauley, of 29-17 Twelfth Street, Long Island City, a deckhand, who was with the East River Ferry Company when the city took over the line sixteen years ago.

On the last run from Manhattan to Astoria, taking the trip for no sentiment, whatever, was Terrence O'Brien, who is a revenue collector for the Department of Plant and Structures. He came to gather up the cash registers at each ferryhouse. On the Manhattan side he got $50, and in Astoria about the same amount.

Behind the quaint headline, "52 Old Friends Bid Good By to Astoria Ferry," were the political realities of how the city would be reshaped and for whose benefit. The new Triboro Bridge would serve a middle-class population that used the car to commute to work. The ferry primarily served the working class, men and women who lived or worked near the ferry terminal.

The most important aspect of the story of the shut-down of the Astoria ferry was the involvement of Robert Moses, whose Triboro Bridge Authority was building the bridge. He wanted the site of the ferry landing at 92nd Street for one of the approach roads to the bridge. A week before the final voyage described in this article, and without advance notice, Moses sent his men to shut down the ferry. As soon as the boat left Manhattan one afternoon for the short trip to Astoria, the men set to work dismantling the slip. On the return trip, the ferry captain had no place to dock. It took Mayor Fiorello LaGuardia's personal intervention, backed by armed policemen, to force the men to stop their work and rebuild the slip. The confrontation proved futile, however, as city commissioners voted to cease operating the ferry within a week anyway.

While Robert Moses did build highways and bridges that benefitted the growing number of car owners, some historians think he lacked concern for those who used public transportation or for communities hurt by his plans.

THE LONG ISLAND CITY-34TH STREET FERRY. "A sad little group of men, grown gray in the service, stood on her after deck watching the wharves and the light-studded buildings recede. One doffed his hat and waved a farewell with the words, 'Good-bye, good-bye.'

"Somewhere in the dusk and distance that were hooding the ferryhouse, a lone, unseen bell was tolling a knell." (March 4, 1925, Daily Star).

QUESTIONS: IDEAS TO THINK ABOUT

1. *How did the ferry link the communities on either side of the river? Give specific examples.*
2. *How did a ferry influence where people chose to live? What was the effect of the ferry on Astoria's growth?*
3. *What did it mean to the community when the ferry was shut down?*
4. *Does the city have an obligation to operate such facilities as the ferry even if it loses money? Why or why not?*
5. *Would it have been possible for the ferry to coexist with the new bridge? What is your opinion of the way Robert Moses tried to close the ferry?*

CROSSING BROOKLYN FERRY

by Walt Whitman

Commuters who must ride the ferry every day are concerned with schedules, the efficiency and convenience of the service. Poets, on the other hand, respond to the experience of riding the ferry differently and come away with other, more sensitive impressions of land and water, of city and nature.

The poet Walt Whitman, inspired by the Brooklyn Ferry as Hart Crane was later inspired by the Brooklyn Bridge, presents some of the images, ideas and feelings communicated to him while "Crossing Brooklyn Ferry."

CROSSING BROOKLYN FERRY

Flood-tide below me! I see you face to face!
Clouds of the west—sun there half an hour high—I see you also face to face.

Crowds of men and women attired in the usual costumes,
how curious you are to me!
On the ferry-boats the hundreds and hundreds that cross returning home,
are more curious to me than you suppose,
And you that shall cross from shore to shore years hence are more to me,
and more in my meditations, than you might suppose.

The impalpable sustenance of me from all things at all hours of the day,
The simple, compact, well-join'd scheme, myself disintegrated,
every one disintegrated yet part of the scheme,
The similitudes of the past and those of the future,
The glories strung like beads on my smallest sights and hearings,
on the walk in the street and the passage over the river,
The current rushing so swiftly and swimming with me far away,
The others that are to follow me, the ties between me and them,
the certainty of others, the life, love, sight, hearing of others.

Others will enter the gates of the ferry and cross from shore to shore,
Others will watch the run of the flood-tide,
Others will see the shipping of Manhattan north and west,
and the heights of Brooklyn to the south and east,
Others will see the islands large and small;
Fifty years hence, others will see them as they cross, the sun half an hour high,
A hundred years hence, or ever so many hundred years hence,
others will see them,
Will enjoy the sunset, the pouring-in of the flood-tide,
the falling-back to the sea of the ebb-tide.

It avails not, time nor place—distance avails not,
I am with you, you men and women of a generation,
or ever so many generations hence,
Just as you feel when you look on the river and sky, so I felt,
Just as any of you is one of a living crowd, I was one of a crowd,
Just as you are refresh'd by the gladness of the river and the bright flow,
I was refresh'd,
Just as you stand and lean on the rail, yet hurry with the swift current,
I stood yet was hurried,
Just as you look on the numberless masts of ships
and the thick-stemm'd pipes of steamboats, I look'd.

I too many and many a time cross'd the river of old,
Watched the Twelfth-month sea-gulls, saw them high in the air
floating with motionless wings, oscillating their bodies,
Saw how the glistening yellow lit up parts of their bodies
and left the rest in strong shadow,
Saw the slow-wheeling circles and the gradual edging toward the south,
Saw the reflection of the summer sky in the water,
Had my eyes dazzled by the shimmering track of beams,
Look'd at the fine centrifugal spokes of light round the shape of my head
in the sunlit water,
Look'd on the haze on the hills southward and south-westward,
Look'd on the vapor as it flew in fleeces tinged with violet,
Look'd toward the lower bay to notice the vessels arriving,
Saw their approach, saw aboard those that were near me,
Saw the white sails of schooners and sloops, saw the ships at anchor,
The sailors at work in the rigging or out astride the spars,

The round masts, the swinging motion of the hulls,
the slender serpentine pennants,
The large and small steamers in motion, the pilots in their pilot-houses,
The white wake left by the passage, the quick tremulous whirl of the wheels,
The flags of all nations, the falling of them at sunset,
The scallop-edged waves in the twilight, the ladled cups,
the frolicsome crests and glistening,
The stretch afar growing dimmer and dimmer,
the gray walls of the granite storehouses by the docks,
On the river the shadowy group, the big steam-tug closely flank'd
on each side by the barges, the hay-boat, the belated lighter,
On the neighboring shore the fires from the foundry chimneys
burning high and glaringly into the night,
Casting their flicker of black contrasted with wild red and yellow light
over the tops of houses, and down into the clefts of streets.

Walt Whitman

QUESTIONS: IDEAS TO THINK ABOUT

1. *What does Walt Whitman suggest about the meaning of Brooklyn Ferry?*
2. *What images does he use in the poem?*
3. *Whitman wrote these lines in the 19th Century. Would New York harbor be unfamiliar to him today?*

FERRIES AND BRIDGES

For every step "Progress" takes, for every technological advance making travel faster, more convenient and cheaper, there are losers as well as winners. While the Queensboro and Triboro Bridges dramatized Queens' leap into the future, they also pushed the ferries in Astoria and Long Island City into the past.

As this article from the Long Island Star-Journal (January 9, 1939) explains, the new Whitestone Bridge sounded the death knell for the College Point Ferry. It points out the relationship between the ferry and other forms of transportation, the automobile and the railroad, in addition to the bridge.

BRONX SPAN DOOMS LINE TO OBLIVION

Route Was City's Busiest Till Tri-Boro Bridge Was Built.

Days are numbered for another old Long Island institution—the venerable College Point ferry which plies between Fourteenth avenue and Flushing Bay and Clason Point in the Bronx.

On April 25, or as soon thereafter as the Whitestone-Bronx Bridge is opened for traffic, the old ferry will be abandoned, probably forever, and thus will come to an end a form of transportation that has been closely identified with Queens and Nassau history for half a century.

The Board of Estimate, meeting yesterday afternoon, adopted a resolution introduced by Comptroller Joseph P. McGoldrick, which sealed the ferry's doom. The action was based on an announcement by the Tri-Borough Bridge Authority, which is constructing the new Whitestone-Bronx span, that the bridge would be thrown open to traffic on the 25th, five days before the opening of the World Fair in Flushing Meadow Park.

Closed Down Before.

The old ferry line, which originally operated from its present terminus to Ninety-ninth street in Manhattan, survived the inroads of two bridges, but could not weather a third span over the East River, almost under its nose.

Its half century of history has not been one of uninterrupted operation. When the Queensboro Bridge, from Long Island City to Fifty-ninth street, opened in 1909, the ferry line faded and went into a decline.

Service ceased after a few months, but another company began operations at a new stand, at the foot of old Tenth street, College Point, where Chisholm Park is now, to Clason Point in the Bronx.

The growth of the Bronx, and increased automobile travel to Westchester, Connecticut and New England points, supported the line for a time, but it, too, faltered and then stopped running.

Then the city, through the Department of Plant and Structures, stepped in, and began service from the present terminal to Clason Point.

Showed a Profit

The ferry prospered—indeed, it became the only line in the city to show a profit, but when business was booming, the second blow landed.

The Tri-Borough Bridge, constructed with a huge federal loan and grant, was opened in 1934, and the College Point ferry's revenue began to drop.

Much of its holiday traffic—remember those mile-long lines of homeward bound cars of a Summer Sunday?—was diverted to the new bridge, although the small army of commuters going from homes in Queens to jobs in the Bronx or vice-versa, remained loyal for the most part.

Service was curtailed as patronage dropped. When plans were announced for the new Whitestone bridge, spanning the East River from the foot of 147th street, it was agreed that the ferry's days were numbered.

Recalls Steamboat Service.

From the earliest days of the steamboat up until only a few years ago, there was regular steamer service from Flushing to Manhattan, with the terminus at the foot of Wall street, but the old College Point-Ninety-ninth street ferry was established as a convenience for farmers.

Around 1895, the Long Island Railroad inaugurated a "farmers' special," consisting of a passenger car, a number of flat cars for farm wagons, and several box cars for the horses. The farmers would load their wagons at various points in Suffolk and Nassau Counties, unload at Jamaica and make for the ferry, which took them to Ninety-ninth street, Manhattan, near the famous old Harlem Market.

The coming of macadam roads and the automobile wiped out this business.

THE BRONX-WHITESTONE BRIDGE UNDER CONSTRUCTION, c. 1939.

QUESTIONS: IDEAS TO THINK ABOUT

1. *What is lost when a bridge replaces a ferry?*
2. *What happened to the ferry pilots and deckhands? In general, what happens when workers are replaced by new technology?*
3. *Did the ferry serve the same public as the bridge would? How would it have been possible for the ferry to continue running?*

THE TROLLEY

The earliest form of mass transit in Queens was the horsecar, forerunner of the electric trolley. This new form of transportation led to a fundamental change in people's lives. For the first time transportation was available to them, if they could afford to use it, rather than their relying on their own resources for travel between villages or from their homes to the ferry.

Imagine how this service must have affected the lives of Queens' residents. The horsecar meant they could travel to Manhattan for shopping, entertainment or work. Of course, if one missed the last horsecar at night it meant a long walk home along dark roads.

The small villages of Queens were connected by an extensive network of private trolley lines, and with a transfer people could travel anywhere in the county for a nickel. Between the towns were open country and farms. For city dwellers, a popular pastime on weekends was taking the Lutheran line out to the cemeteries in Middle Village. Paying respect at the grave of a loved one was part of a pleasant ride through the countryside and a picnic in the cemetery. The horse car line from Borden Avenue to Calvary Cemetery began operation on March 1, 1874, and like the other lines in Queens, was electrified in the 1890's.

New Advertisements.

SPECIAL NOTICE.

THE
ASTORIA AND HUNTERS POINT
RAILROAD COMPANY
Will Commence Running their Cars
THROUGH TO ASTORIA,
On Saturday, June 19, 1869.
On that day there will be no charge.
Let the Public enjoy a
FREE RIDE
OVER THE ROAD.

Notice will hereafter be given with regard to fare and the time of starting from each point.
June 16, 1869.
C. R. TRAFFORD,
President.

Pictured here in front of Tony Miller's Long Island City Hotel on Borden Avenue is the horse car which ran along Vernon Avenue to Astoria. The advertisement announces that on the first day of operation, June 19, 1869, the public could enjoy a "Free Ride over the road" to introduce them to the new service.

QUESTIONS: IDEAS TO THINK ABOUT

1. *What had to happen before trolleys could replace horsecars? What was the meaning of that change for the drivers? The riders?*
2. *What is your impression of Long Island City from these photographs?*

BORDEN AVENUE NEAR THE 34TH STREET FERRY, 1896. Tony Miller's Hotel is in the background.

HORSECARS TO NORTH BEACH

From the ferry landings people took trolleys that travelled the local streets of Long Island City and Astoria. The most popular line, however, was the one that ran out Astoria Boulevard to North Beach, a popular resort area opened on the shore of Bowery Bay in 1886.

As the following article explains, North Beach was developed by George Ehret, one of New York's most successful brewers, William Steinway, the piano manufacturer and owner of a horsecar line, and J. L. Mott, the "iron king." These friends and partners prospered from this arrangement where people rode Steinway's horsecars along Mott's rails to restaurants and beer gardens where they drank Ehret's beer. Many political clubs, workers' associations, and ethnic societies went on outings to North Beach. For an all inclusive price they enjoyed transportation by trolley, food, beer, and entertainment.

North Beach went into decline when prohibition became law in 1919; a fire in 1924 destroyed many of the buildings. Later, the site was developed as New York's second municipal airport, opened as LaGuardia Airport in 1939.

QUESTIONS: *IDEAS TO THINK ABOUT*

1. Was it ethical for Ehret, Steinway, and Mott to cooperate in this venture? How would their partnership have limited competition?

2. Who patronized North Beach? Why did they travel there? Describe how you imagine an outing to North Beach would have been.

3. What was lost when Prohibition forced establishments in North Beach to close? What did people do for leisure instead?

JAMAICA AVENUE AND 160TH STREET, 1914.

Jamaica, like Long Island City, was a transportation hub. The railroad had its busiest season there, and trolley lines from Brooklyn, Queens and Nassau converged on Jamaica Ave. Because so many people passed through the area every day, it was a natural location for many kinds of businesses. In this 1914 photograph we see several restaurants and saloons, a shoe-shine stand, and even a paper boy.

QUESTIONS: IDEAS TO THINK ABOUT

1. *What details here give you a sense of what life was like in 1914?*
2. *What ethnic groups are in evidence here?*
3. *How would this scene look today?*

CHAMBER OF COMMERCE OF THE BOROUGH OF QUEENS. 43

SIDE DOOR CAR OPERATED BY MANHATTAN & QUEENS TRACTION CORPORATION ON QUEENS BOULEVARD; STATION ON ELEVATED RAILROAD, SHOWING CONCRETE FORM OF CONSTRUCTION. PHOTO TAKEN FEBRUARY, 1915.

The following table shows the growth of street railway traffic on lines operating in Queens from 1898 to 1914:

Year	*No. of Fares Collected*	*Annual Increase*	*Per Cent Increase*
1898	9,128,804	....	..
1899	9,472,460	343,656	3.7
1900	11,441,751	1,969,291	20.8
1901	11,564,062	122,311	1.1
1902	13,564,062	2,155,325	18.6
1903	15,689,210	1,969,823	14.4
1904	16,701,653	1,012,443	6.5
1905	20,533,487	3,831,834	22.9
1906	25,151,054	4,617,567	22.5
1907	28,514,743	3,363,689	13.4
1908	29,797,750	1,283,007	4.5
1909	30,545,776	748,026	2.5
1910	34,430,074	3,884,298	12.7
1911	42,522,485	8,092,411	23.4
1912	45,182,732	2,660,247	6.2
1913	47,463,382	2,280,650	5.0
1914	49,973,696	2,510,314	5.3

The following table gives the traffic on the various trolley companies operating all or in part of the Borough of Queens, for the past three years, but does not include the traffic on the various trolley lines operated by the Brooklyn Rapid Transit Co., such as the line on Jamaica Avenue from Cypress Hills to Jamaica, which carries over a million passengers a month:

Railroad	*Length of Route (Miles)*	*No. of Fare Passengers for Year Ending June 30* 1912	1913	1914
1—New York & Queens Co. Ry.	75.35	25,450,728	26,950,656	26,744,147
2—New York & Long Island Tract. Co.	41.31	7,834,539	8,040,320	8,088,288
3—Long Island Elec. Ry. Co.	26.56	3,826,087	4,084,666	4,367,692
4—New York & North Shore Tract. Co.	37.68	2,755,140	2,761,466	2,878,546
5—Ocean Elec. Ry. Co.	8.59	2,647,904	2,876,607	3,171,551

This page from a 1915 publication of the Queensborough Chamber of Commerce gives various statistics on trolley lines in Queens. By reading this document you see not only the number of fares collected and the number of miles of track, but also how Queens' population grew from 1898 to 1915.

INCIDENT
by Countee Cullen

On the preceding pages we have looked at the impact trolleys had on communities. We have seen statistics showing the annual increase in trolley ridership, but no indication of who those people were, nor any mention of the new kinds of interpersonal problems created when individuals of different backgrounds and temperament have to use the same facilities.

Riding mass transportation—trolleys, buses, subways, and ferries—forces us to come in contact with different kinds of people with different habits. We must daily confront our own prejudices and must stretch the limits of our tolerance. This short but touching poem describes the contact between two children, one black and the other white, on a trolley in Baltimore, where at one time all public transportation was segregated.

Incident
(For Eric Walrond)

Once riding in old Baltimore,
Heart-filled, head-filled with glee,
I saw a Baltimorean
Keep looking straight at me.

Now I was eight and very small,
And he was no whit bigger,
And so I smiled, but he poked out
His tongue, and called me, "Nigger."

I saw the whole of Baltimore
From May until December;
Of all the things that happened there
That's all that I remember.

Countee Cullen (1903-1946)

QUESTIONS: IDEAS TO THINK ABOUT

1. *Have you seen or experienced an incident of prejudice or cruelty on public transportation? Describe it and the people involved.*
2. *What kinds of behavior on public transportation are, or should be prohibited? Why? How do you feel when behavior codes are broken?*

As the trolley became an increasingly important feature of the urban scene, trolley workers also assumed greater importance. In 1895, the workers went on strike against the Brooklyn Rapid Transit Co., which owned almost all of the trolley lines in Brooklyn. The men were paid $2 for a 12 hour day at the time, and went on strike for a shorter workday. During the bitter two week strike, the company brought in scab labor from other cities to keep the cars running.

When the cars were pulled by horses, their maximum speed was only 5-7 miles per hour, and the drivers on the longer runs made only one circuit a day. With electrification, they made twice as many trips as before. In a sense, the strike was a response by workers to new technology which doubled the amount of work required.

MAINTENANCE CREW IN THE MASPETH CAR BARNS, 1927.

During the strike the workers effectively shut down the entire system. They put logs on the tracks and pieces of metal wire to snag the guide wires. The governor called out the National Guard which camped on Alabama Avenue in East New York where the car barns were located. Each day during the strike, one trolley car crowded with soldiers made its way through streets lined with jeering strikers. The company was able to say that the cars were running, but no one rode except the soldiers. At the end of two weeks, however, nearly all the trolley lines were operating again and the strike was broken. Less than half of the workers who struck got their jobs back, and those who did had to start at the bottom.

QUESTIONS: IDEAS TO THINK ABOUT

1. *What new occupations were created by trolley travel?*
2. *What are the usual issues when workers strike? Were the issues in this trolley strike similar or different?*
3. *Should transit workers, or any public employees, be permitted to strike? How are the concerns of the workers to be weighed with the rights of the public?*
4. *In 1981, the Air Traffic Controllers went on strike. How were the issues in that strike similar to those in the trolley strike described here?*

SISTER CARRIE

by Theodore Dreiser

In *Sister Carrie*, a novel written in 1900, Theodore Dreiser describes such a trolley strike as the one in Brooklyn. He focuses on the experience of Hurstwood, Carrie's unemployed husband, who worked as a scab driver for a day. In this passage we get a sense of the violent emotions running on both sides, and we also hear the class consciousness of the strikers as they try to explain their point of view to the scab drivers before the policemen escorting the trolley lash out with their nightsticks.

SISTER CARRIE

Hurstwood climbed up on the platform in the gloomy barn and waited for a signal. He was nervous, and yet the thing was a relief. Anything was better than the barn.

On this the fourth day of the strike, the situation had taken a turn for the worse. The strikers, following the counsel of their leaders and the newspapers, had struggled peaceably enough. There had been no great violence done. Cars had been stopped, it is true, and the men argued with. Some crews had been won over and led away, some windows broken, some jeering and yelling done; but in no more than five or six instances had men been seriously injured. These by crowds whose acts the leaders disclaimed.

Idleness, however, and the sight of the company, backed by the police, triumphing, angered the men. They saw that each day more cars were going on, each day more declarations were being made by the company officials that the effective opposition of the strikers was broken. This put desperate thoughts in the minds of the men. Peaceful methods meant, they saw, that the companies would soon run all their cars and those who had complained would be forgotten. There was nothing so helpful to the companies as peaceful methods.

All at once they blazed forth, and for a week there was storm and stress. Cars were assailed, men attacked, policemen struggled with, tracks torn up, and shots fired, until at last street fights and mob movements became frequent, and the city was invested with militia.

Hurstwood knew nothing of the change of temper.

"Run your car out." called the foreman, waving a vigorous hand at him. A green conductor jumped up behind and rang the bell twice as a signal to start. Hurstwood turned the lever and ran the car out through the door into the street in front of the barn. Here two brawny policemen got up beside him on the platform—one on either hand.

At the sound of a gong near the barn door, two bells were given by the conductor and Hurstwood opened his lever.

The two policemen looked about them calmly.

"'Tis cold, all right, this morning," said the one on the left, who possessed a rich brogue.

"I had enough of it yesterday," said the other. "I wouldn't want a steady job of this."

"Nor I."

Neither paid the slightest attention to Hurstwood, who stood facing the cold wind, which was chilling him completely, and thinking of his orders.

"Keep a steady gait," the foreman had said. "Don't stop for any one who doesn't look like a real passenger. Whatever you do, don't stop for a crowd."

The two officers kept silent for a few moments.

"The last man must have gone through all right," said the officer on the left. "I don't see his car anywhere."

"Who's on there?" asked the second officer, referring, of course, to its complement of policemen.

"Schaeffer and Ryan."

There was another silence, in which the car ran smoothly along. There were not so many houses along this part of the way. Hurstwood did not see many people either. The situation was not wholly disagreeable to him. If he were not so cold, he thought he would do well enough.

He was brought out of this feeling by the sudden appearance of a curve ahead, which he had not expected. He shut off the current and did an energetic turn at the brake, but not in time to avoid an unnaturally quick turn. It shook him up and made him feel like making some apologetic remarks, but he refrained.

"You want to look out for them things," said the officer on the left, condescendingly.

"That's right," agreed Hurstwood, shamefacedly.

"There's lots of them on this line," said the officer on the right.

Around the corner a more populated way appeared. One or two pedestrians were in view ahead. A boy coming out of a gate with a tin milk bucket gave Hurstwood his first objectionable greeting.

"Scab!" he yelled. "Scab!"

Hurstwood heard it, but tried to make no comment, even to himself. He knew he would get that, and much more of the same sort, probably.

At a corner farther up a man stood by the track and signalled the car to stop.

"Never mind him," said one of the officers. "He's up to some game."

Hurstwood obeyed. At the corner he saw the wisdom of it. No sooner did the man perceive the intention to ignore him, than he shook his fist.

"Ah, you bloody coward!" he yelled.

Some half dozen men, standing on the corner, flung taunts and jeers after the speeding car.

Hurstwood winced the least bit. The real thing was slightly worse than the thoughts of it had been.

Now came in sight, three or four blocks farther on, a heap of something on the track.

"They've been at work, here, all right," said one of the policemen.

"We'll have an argument, maybe," said the other.

Hurstwood ran the car close and stopped. He had not done so wholly, however, before a crowd gathered about. It was composed of ex-motormen and conductors in part, with a sprinkling of friends and sympathisers.

"Come off the car, pardner," said one of the men in a voice meant to be conciliatory. "You don't want to take the bread out of another man's mouth, do you?"

Hurstwood held to his brake and lever, pale and very uncertain what to do.

"Stand back," yelled one of the officers, leaning over the platform railing. "Clear out of this, now. Give the man a chance to do his work."

"Listen, pardner," said the leader, ignoring the policeman and addressing Hurstwood. "We're all working men, like yourself. If you were a regular motorman, and had been treated as we've been, you wouldn't want any one to come in and take your place, would you? You wouldn't want any one to do you out of your chance to get your rights, would you?"

"Shut her off! shut her off!" urged the other of the policemen, roughly. "Get out of this, now," and he jumped the railing and landed before the crowd and began shoving. Instantly the other officer was down beside him.

"Stand back, now," they yelled. "Get out of this. What the hell do you mean? Out, now."

It was like a small swarm of bees.

"Don't shove me," said one of the strikers, determinedly. "I'm not doing anything."

"Get out of this!" cried the officer, swinging his club. "I'll give ye a bat on the sconce. Back, now."

"What the hell!" cried another of the strikers, pushing the other way, adding at the same time some lusty oaths.

Crack came an officer's club on his forehead. He blinked his eyes blindly a few times, wabbled on his legs, threw up his hands, and staggered back. In return, a swift fist landed on the officer's neck.

Infuriated by this, the latter plunged left and right, laying about madly with his club. He was ably assisted by his brother of the blue, who poured ponderous oaths upon the troubled waters. No severe damage was done, owing to the agility of the strikers in keeping out of reach. They stood about the sidewalk now and jeered.

"Where is the conductor?" yelled one of the officers, getting his eye on that individual, who had come nervously forward to stand by Hurstwood. The latter had stood gazing upon the scene with more astonishment than fear.

"Why don't you come down here and get these stones off the track?" inquired the officer. "What you standing there for? Do you want to stay here all day? Get down."

Hurstwood breathed heavily in excitement and jumped down with the nervous conductor as if he had been called.

"Hurry up, now," said the other policeman.

Cold as it was, these officers were hot and mad. Hurstwood worked with the conductor, lifting stone after stone and warming himself by the work.

"Ah, you scab, you!" yelled the crowd. "You coward! Steal a man's job, will you? Rob the poor, will you, you thief? We'll get you yet, now. Wait."

Not all of this was delivered by one man. It came from here and there, incorporated with much more of the same sort and curses.

"Work, you blackguards," yelled a voice. "Do the dirty work. You're the suckers that keep the poor people down!"

"May God starve ye yet," yelled an old Irish woman, who now threw open a nearby window and stuck out her head.

"Yes, and you," she added, catching the eye of one of the policemen. "You bloody, murtherin' thafe! Crack my son over the head, will you, you hardhearted, murtherin' divil? Ah, ye——"

But the officer turned a deaf ear.

"Go to the devil, you old hag," he half muttered as he stared round upon the scattered company.

Now the stones were off, and Hurstwood took his place again amid a continued chorus of epithets. Both officers got up beside him and the conductor rang the bell, when, bang! bang! through window and door came rocks and stones. One narrowly grazed Hurstwood's head. Another shattered the window behind.

"Throw open your lever," yelled one of the officers, grabbing at the handle himself.

Hurstwood complied and the car shot away, followed by a rattle of stones and a rain of curses.

"That — — — —— hit me in the neck," said one of the officers. "I gave him a good crack for it, though."

"I think I must have left spots on some of them," said the other.

"I know that big guy that called us a — — — ——," said the first. "I'll get him yet for that."

"I thought we were in for it sure, once there," said the second.

Hurstwood, warmed and excited, gazed steadily ahead. It was an astonishing experience for him. He had read of these things, but the reality seemed something altogether new. He was no coward in spirit. The fact that he had suffered this much now rather operated to arouse a stolid determination to stick it out. He did not recur in thought to New York or the flat. This one trip seemed a consuming thing.

Theodore Dreiser

QUESTIONS: IDEAS TO THINK ABOUT

1. *What are the interests of the strikers in this passage? The police? Hurstwood?*
2. *Do you know anyone who worked as a strikebreaker, a "scab"? What was his or her rationale for doing so?*

WOMEN TROLLEY WORKERS

These photographs show that women as well as men worked on the trolley lines. It is interesting, however, that the pictures were taken during the First and Second World Wars when there were shortages of male workers. One picture taken in 1918 shows a group of women conductors in the car barn of the New York and Queens County Railway Co., in Woodside. The second photo shows a driver about to begin her run from the Maspeth car barn to Flushing in 1945. One wonders what happened to these women after the war, whether they kept their jobs or lost them to returning veterans.

TROLLEY CONDUCTORS, WOODSIDE CAR BARN, 1918.

TROLLEY DRIVER, MASPETH CAR BARN, 1945.

QUESTIONS: IDEAS TO THINK ABOUT

1. *What do you think happened to these women after the wars were over? How did they feel?*
2. *Why are some jobs assigned only to men while others are assigned exclusively to women?*

TROLLEY ACCIDENT, c.1935

This tragic accident occurred on the right of way of the Queens County Railway Co. in Jamaica, just south of Hillside Ave. There is an unreal quality to the scene, as a boy's body lies on the tracks with no one near. A pair of newsmen have set up their camera on the left, and a crowd of trolley officials and police officers are gathered around the trolley cars further down the alley. A woman stares at the scene from her rooftop.

People who lived along the trolley line worried about safety and the noise at night as the trolley rumbled by. While new transportation systems did provide a service to the community, they also caused new problems. Sometimes there was conflict between the trolley company putting in a new line and the neighborhood that would feel the bad effects of "progress" rather than its benefits.

This tragic accident occurred on the tracks of the Queens County Railway Co. in Jamaica, just south of Hillside Ave., 1935.

QUESTIONS: IDEAS TO THINK ABOUT

1. *Have you ever witnessed an accident involving public transportation? Describe what happened.*
2. *In what ways has transportation safety improved over the years? What should have been done to prevent the accident pictured here?*

THE LAST TROLLEY

The Queensboro Bridge was designed to carry trolleys as well as automobiles and wagons. In 1909 when the bridge opened, it had four tracks, two on the outer roadways and two more on the main road. The trolley began simply as a shuttle, with a fare of 2 1/2¢ each way over the bridge. By 1912 the trolley was connected to the other lines in Bridge Plaza, and eventually riders could take the trolley from Manhattan all the way down Queens Boulevard to Jamaica.

The trolley on the outer roadway, pictured here shortly after it began running in 1909, ended its service in 1957 as a shuttle. It was the last trolley in regular operation in all of New York State.

QUESTIONS: IDEAS TO THINK ABOUT

1. *Why did New York City decide to remove trolleys from operation? How did this decision affect the communities along the trolley routes?*
2. *Today, the Q-60 bus follows the route of the trolley which ran over the Queensboro Bridge and down Queens Blvd. to Jamaica. How does the bus compare with the trolley in terms of efficiency, comfort, or convenience?*

THE BRIDGE

For Queens, the new Queensboro Bridge dramatized the change from the nineteenth century to the twentieth. It meant modern technology, new people, and growth occurring at a dizzying pace. New York's largest borough exploded in population and industry. Within ten years of the opening of the bridge in 1909, the surrounding landscape of swamps and farms became a major industrial center. The steel beams of the world's largest cantilever bridge eventually carried people in trolleys, in horse drawn wagons, in automobiles, in the elevated trains, and by foot across the East River.

Poets and artists view the bridge as the leap of the imagination, a challenge to temporary limitations on humanity's capacities. Yet businessmen, merchants, or workers see something else in the bridge: it means new opportunities, new investments, new jobs. The Queensboro Bridge has fulfilled the dreams and visions of both viewpoints.

THE BRIDGE

by Hart Crane

Perhaps the most famous poem about bridges is Hart Crane's "The Bridge," inspired by the Brooklyn Bridge. Crane wrote the poem in the room where Washington Roebling, the paralyzed engineer of the Brooklyn Bridge, sat and observed the construction of it. Crane said of his 62 page poem that "...it is based on the conquest of space and knowledge. ... The Bridge becomes the symbol of consciousness spanning time and space."

Below are a few stanzas from the beginning of the poem.

THE BRIDGE

And thee, across the harbor, silver-paced
As though the sun took step of thee, yet left
Some motion ever unspent in thy stride,—
Implicitly thy freedom staying thee!

Out of some subway scuttle, cell or loft
A bedlamite speeds to thy parapets,
Tilting there momently, shrill shirt ballooning,
A jest falls from the speechless caravan.

Down wall, from girder into street noon leaks,
A rip-tooth of the sky's acetylene;
All afternoon the cloud-flown derricks turn...
They cables breathe the North Atlantic still.

O harp and altar, of the fury fused,
(how could mere toil align thy choiring strings!)
Terrific threshold of the prophet's pledge,
Prayer of pariah, and the lover's cry,

Again the traffic lights that skim thy swift
Unfractioned idiom, immaculate sigh of stars,
Beading thy path—condense eternity:
And we have seen night lifted in thine arms.

Under thy shadow by the piers I waited;
Only in darkness in thy shadow clear.
The city's fiery parcels all undone,
Already snow submerges an iron year...

O sleepless as the river under thee,
Vaulting the sea, the prairies' dreaming sod,
Unto us lowliest sometime sweep, descend
And of the curveship lend a myth to God.

Hart Crane (1930)

QUESTIONS: IDEAS TO THINK ABOUT

1. *What does a bridge symbolize for writer and artist?*
2. *Compare this poem with "Crossing Brooklyn Ferry." Are the images and the meaning the same?*

QUEENSBORO BRIDGE UNDER CONSTRUCTION, JANUARY 9, 1907.

This picture of the cantilever section extended over the east channel of the river shows how impressive this technological feat must have seemed. Yet for the men who built the bridge it was a dangerous job; fifty men lost their lives in the nine years it took to complete work.

The bridge was built by two rival crews, one constructing the span over the west channel between Blackwell's Island and Manhattan, and the other the span over the east Channel to Queens. When the Manhattan gang, led by Jim Heddrick, finished their task they celebrated against the skyline, among the girders high above Manhattan's east side.

When I was a boy, say in 1903, my grandfather used to take me, then I was about five years old, to the different farms that was left. We were going there one day ... now this is bringing back a lot of memories; we were going to Nolte's farm. ... I happened to look over and saw the Queensboro Bridge, standing on Blackwell's Island as it was called then ... like something standing on stilts. I said to grandpa, I says "Grandpa, what is that? What's it doing? What's it for?" There was nothing over the water. It was just standing right on Blackwell's Island.

George Steinmayer

QUESTIONS: IDEAS TO THINK ABOUT

1. *Has the building of a bridge ever affected your life?*
2. *Have you ever watched a structure like the bridge being built over several years? What was the impact on your neighborhood?*
3. *How do our personal memories compare with the actual historical events or process? How can those memories add to our understanding of history?*

THE OPENING OF THE BRIDGE, MARCH 30, 1909.

After almost ten years and a cost of $18 million, the Queensboro Bridge was finally opened to traffic. Mayor McClellan leads a procession of cars from the Manhattan side. The tolls announced on the sign, 3¢ for "horses without vehicles or on lead" and 10¢ for automobiles, remained in effect until July, 1911 when they were abolished by the city aldermen. The rates do not seem so low when you consider that in 1915 the subway only cost 5¢, and that people working in factories earned about $5 a week. If they could not walk to work, transportation cost them about one fifth of their wages.

QUESTIONS: IDEAS TO THINK ABOUT

1. *Why were the tolls abolished after two years?*
2. *Should tolls be reinstated on the Queensboro Bridge?*

BRIDGE PLAZA, 1909.

On the Queens side of the Bridge was a spacious, well designed plaza with a small park in the center, around which the trolley made its turn for the trip back to New York. The flagpole was a gift from Commodore Lipton, the tea merchant. The most prominent building is the home of the Brewster Co., the first automobile company to locate in Long Island City and where George Steinmayer, the boy who watched the construction of the bridge with his grandfather, got a job in 1919.

Designed like a village green, the open and attractive space is nonetheless deceiving. The plaza was layed out with the intention of building a large station for elevated trains that would extend the city's transit system into Queens. The bridge was designed for trains as well as trolleys, cars, wagons and pedestrians; within five years, two elevated lines converged at this plaza.

Looking towards Bridge Plaza only eight years later, note the incredible changes. The plaza where a single trolley had made its graceful loop around the green was now cluttered with the new subway lines, the Second Avenue El, the tracks of which are seen here on the upper roadway with a pedestrian walkway on either side, and the IRT Flushing line, extending in a line from the right. The distinctive clocktower of the Brewster Building still looms over the horizon, but this building no longer reigns supreme as the only large edifice against the sky. These new factories provided hundreds of new jobs, spurring the growth of nearby communities like Astoria.

LOOKING EAST TOWARDS BRIDGE PLAZA, 1917.

QUESTIONS: IDEAS TO THINK ABOUT

1. *What happened to nearby land values when the bridge was built? Who made a profit?*
2. *Why are factories and banks still located in Bridge Plaza? How does that affect the neighborhood?*

THE QUEENSBOROUGH BRIDGE, EAST RIVER
by Glenn Coleman, 1910

In Glen Coleman's oil painting the bridge seems to be part of the New York scene. The people in this painting accept the bridge with casual indifference. Like the gas tanks and factories across the East River, the excursion boat steaming upriver, or the laundry hung out to dry on a clothesline, the Queensborough Bridge here is part of the urban landscape. On one level bridges represent a people's conquest over a natural barrier, the victory of technology over nature, but for many artists and writers, bridges have a special beauty: they reach to the sky, and express the limitlessness of human potential and ambition.

QUESTIONS: IDEAS TO THINK ABOUT

1. *What feeling does the painting give you?*
2. *Do you know of other artistic representations of city landmarks?*

ALEXANDER'S BRIDGE
by Willa Cather

In *Alexander's Bridge,* a novel published in 1912, readers followed an actual event, the collapse in 1909 of a new bridge under construction across the St. Lawrence River. Cather was living in New York City during the construction of the Queensborough Bridge and perhaps she watched its steel girders grow and used some of her observations when she described the fictional Moorlock Bridge.

The novel concerns Bartley Alexander, a renowned American bridge builder in his middle forties who is constructing his most important bridge in Canada. This bridge is to be the "longest cantilever in existence."

Alexander is an extremely successful man with the vision that "builds the bridges into the future, over which the feet of everyone of us will go." He has a beautiful wife whom he loves very much. However, on a trip to London, he resumes a love affair with a beautiful actress he had known intimately in his youth. He finds himself torn between these two women. Unable to decide what to do, he does nothing; his indecision causes him to neglect his bridge in Canada. He is suddenly called back from Europe to his bridge because strains develop in the half-completed bridge. Finally, before his very eyes, the bridge collapses:

> "Once the chords began to buckle, there were thousands of tons of ironwork all riveted together and lying in mid-air without support. It tore itself to pieces with roaring and grinding and noises that were like the shrieks of a steam whistle. . . . It lurched neither to right nor left, but sank almost in a vertical line, snapping and breaking, as it went and tearing because no integral part could bear for an instant the enormous strain loosed upon it."

QUESTIONS: IDEAS TO THINK ABOUT

1. *Building a bridge represents technology conquering nature. Do we pay a price for this? How? Is it worth it?*

Bridges demonstrate in a most dramatic way the triumph of man's vision and skill over a natural barrier. They represent the movement from the past to the future, a direct route from the old neighborhood to the new, from the known and familiar to the unknown and challenging. Carolyn M. Rodgers uses the symbol of a bridge to represent her personal connection to her mother and to her Black roots in her poem, "It is deep."

IT IS DEEP

(don't never forget the bridge that you crossed over on)

Having tried to use the
witch cord
that erases the stretch of
thirty-three blocks
and turning in the voice which
woodenly stated that the
talk box was "disconnected"

My mother, religiously girdled in
her god, slipped on some love, and
laid on my bell like a truck
blew through my door warm wind from the south
concern making her gruff and tight-lipped
and scared
that her baby was starving
she, having learned, that disconnection results from non-payment
of bill(s).

She did not
recognize the poster of the
grand le-roi (al) cat on the wall
had never even seen the book of
black poems that I have written
thinks that I am under the influence of
communists
when I talk about Black as anything
other than something ugly to kill it before it grows
in any impression she would not be
considered "relevant" or "Black"
but
there she was, standing in my room
not loudly condemning that day and
not remembering that I grew hearing her
curse the factory where she "cut uh slave"
and the cheap J-boss wouldn't allow a union,
not remembering that I heard the tears when
they told her a high school diploma was not enough,
and here now, not able to understand, what she had
been forced to deny, still—

she pushed into my kitchen so
she could open my refrigerator to see
what I had to eat, and pressed fifty
bills in my hand saying "pay the talk bill and buy
some food; you got folks who care about you..."

My mother, religious-negro, proud of
having waded through a storm, is very obviously,
a sturdy Black bridge that I
crossed over, on.

Carolyn M. Rodgers

QUESTIONS: IDEAS TO THINK ABOUT

1. *What does a bridge symbolize for Carolyn M. Rodgers?*
2. *Compare "It is Deep" with "The Bridge" by Hart Crane. Which poem speaks more directly to you?*

FUNERAL PROCESSION CROSSING PENNY BRIDGE, c. 1910, Newtown Creek.

FUNERAL PROCESSION CROSSING PENNY BRIDGE, c. 1910

Not all bridges possess the majesty and technological grace of the Queensborough, nor do all have the same explosive impact. Still, they may give us cause for reflection or inspiration. If a bridge can represent our journey from the past to a less certain future, from a world we know to one unknown, then what are we to make of this photograph by Frederick J. Weber of a funeral procession crossing Newtown Creek over Penny Bridge? The human drama contrasts sharply with the stark industrial landscape, the polluted water, the smoke pouring from the stacks of the oil refinery.

QUESTIONS: IDEAS TO THINK ABOUT

1. *What details in the photograph contribute to its mood?*

THE RAILROAD

Before subways, the main form of transportation for people living on Long Island who wanted to travel to New York (Queens and Brooklyn did not become part of New York City until 1898) was the Long Island Railroad. The railroad had operated under several names and owners since the middle of the nineteenth century, and it was bought by the Pennsylvania Railroad in 1900.

The railroad ran from eastern Long Island through Jamaica to either Atlantic Avenue in downtown Brooklyn or Long Island City. Before the subway was extended to Jamaica, the railroad provided a special "rapid transit" service for people. The train, which ran above ground, made local stops along Atlantic Avenue; the fare was 10¢. When the Atlantic Avenue subway opened in 1908, people could travel all the way to the Nassau County line for 15¢. Within the city however, the railroad proved unable to compete with the less expensive subways subsidized by the city and looked to the growing suburbs of Long Island for most of its business.

COMMUTERS INCREASE ON LONG ISLAND R. R.

Fourteen Per Cent. More Tickets Sold Than at This Time Last Year.

STATION PLANS CHANGED.

More Room Is Required at Big Manhattan Depot to Handle Long Island Traffic.

Apr. 26 1911

The Long Island Railroad Company sold for April this year 11,104 monthly commutation tickets, as compared with 9,748 sold during April of last year, a gain of about 14 per cent. The great movement from Thirty fourth street to the Pennsylvania station has been supplemented by an increase in the total uptown traffic of over 55 per cent. The total by Thirty-fourth Street Ferry, the only route last April, was 2,776. This month the total by Thirty fourth street Ferry was only 812, but the surprising total of 3,413 was reached by Pennsylvania station.

The Long Island business is now increasing so rapidly at that station that the wall at the east end of the Long Island Railroad waiting room in the big station is being knocked out and the space now occupied by one of the corridors will be taken clear through to Seventh avenue for the enlargement of the present waiting room and ticket offices. It is expected that the traffic to the beaches from the big uptown station will be enormous over the electric lines to Rockaway and Long Beach without change of cars at Jamaica.

The number of commuters to Long Island City fell off largely, but there was considerable gain for Flatbush avenue. The total for the two stations in April, 1910, was 6,972. This year it is 5,779. The falling off is due to the very large reduction at Long Island City.

Following is a comparison between the number sold at the principal stations in April of last year and this:

	1911.	1910.
Garden City	276	132
Mineola	84	69
Sea Cliff	186	173
Floral Park	162	117
Jamaica	442	337
Morris Park	220	202
Queens	132	119
Nassau Boulevard	73	9
Huntington	181	162
Hempstead	313	313
Woodside	31	3
Corona	198	139
Flushing	237	230
Murray Hill	438	391
Bay Side	292	288
Port Washington	153	137
Great Neck	138	155
Richmond Hill	200	308
Valley Stream	120	100
Woodmere	132	99
Cedarhurst	147	134
Lawrence	135	140
Far Rockaway	691	647
Arverne	148	139
Rockaway Beach	366	315
Lynbrook	293	291
Long Beach	135	123
Rockville Centre	589	480
Baldwins	143	127
Freeport	532	477
Merrick	102	73
Amityville	174	172
[illegible]	[illegible]	[illegible]

In 1911, the Long Island Railroad looked back on a year when its ridership increased by 14%, and the company anticipated continued success as the population of Long Island grew. The tunnels to Pennsylvania Station completed, the increased number of riders traveling directly into Manhattan more than offset the decline in passengers using the old Long Island City-34th Street ferry route. The railroad had a virtual monopoly on east-west travel on the island.

Within a few years, however, the railroad faced competition from other forms of transportation. Subways took much of the railroad's business within the city, and new highways out to the rapidly growing suburbs would soon make commuting by car easy and desirable.

QUESTIONS: IDEAS TO THINK ABOUT

1. *What were the reasons for the railroad's optimism in 1911?*
2. *The railroad did not anticipate the reasons for its decline. Why not?*

In this advertisement from 1942 the Long Island Railroad explains the need for a fare increase (fares had not changed since 1918) and a reduction in taxes. Part of the complaint was that the owners, the Pennsylvania Railroad, had invested millions of dollars in new equipment and upgrading facilities, but had not made a profit. One reason for this was that the newly opened IND lines on Queens Boulevard took most of the railroad's intra-city commuter business. The company points out that its taxes help pay the operating losses of the subway, while the subway takes passengers from the railroad.

ONCE THIS STATION HAD 2,415* COMMUTERS...

NOW IT HAS 89... *Why?***

THIS STATION is familiar to nearly all Long Island Rail Road commuters — it's the beautiful Forest Hills Station, on the run from Pennsylvania Station to Jamaica.

Its city-bound platform used to be crowded with commuters in rush morning hours. Less than 100 commuters use the station now.

Why? Because of the extension of the city-owned Eighth Avenue Subway through Forest Hills to Jamaica. Even if the railroad were to cut its 60-trip commutation rate in half, it could not meet the 5c subway fare. But a municipality can arbitrarily set a fare for its self-owned lines, regardless of whether that fare meets expenses. Its operation is tax-free and the deficits are made up by the taxpayers.

But a railroad has to make both ends meet and pay its way and its taxes, too. In fact, taxes paid by the Long Island Rail Road help pay the operating losses on subsidized subway operations which take business away from the railroad.

What has happened at Forest Hills has happened wherever city subways have extended.

These Two Charts Show How Subway Competition Has Cut Into Long Island R.R. Passenger Business

BEFORE

BEFORE SUBWAY EXTENSIONS WERE COMPLETED, 43% of Long Island Rail Road commuting patronage came from within a radius of 15 miles of Pennsylvania Station. It was equal to 34,000,000 passengers a year.

AFTER

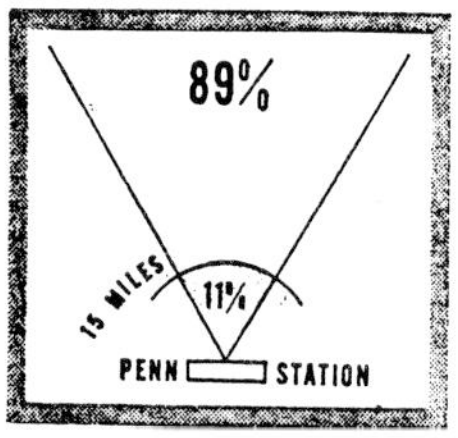

AFTER SUBWAY EXTENSIONS WERE COMPLETED, commuter patronage within that radius fell to about 5,000,000 passengers a year by 1941, or 11% of that year's total; 29,000,000 fewer passengers than in 1931.

But loss of Long Island Rail Road passengers to subway competitors didn't end there. "Feeding" bus lines now carry to subways a large number of former commuters from territory east of the subway terminals and west of Little Neck, Floral Park and Valley Stream. So altogether between May, 1931 and May, 1941, subways have taken 60 PER CENT OF THE RAILROAD'S COMMUTER BUSINESS from the area west of the points mentioned.

The White Report shows that this has resulted in a heavy loss on passenger service operation since operating expenses cannot be reduced in proportion to the decrease in travel. This operating loss on passenger service amounted to $20,640,368 for the seven years ended Dec. 31, 1940. That's why the White Report recommends that the Long Island Rail Road seek an increase in commutation fares (which are exactly where they were in 1918), and a reduction in its taxes (now practically two-and-one-half times what they were twenty years ago), in order to continue efficient service and warrant investment in widespread improvements.

LONG ISLAND RAIL ROAD

BUY UNITED STATES WAR BONDS AND STAMPS

QUESTIONS: IDEAS TO THINK ABOUT

1. *Are the railroad's arguments for a fare increase and tax reduction legitimate? Why or why not? What other problems caused the railroad to lose riders?*
2. *The railroad complains today that it operates at a loss. What are the reasons for that? Should the riders pay all operating costs, or should the railroad be subsidized by the government? Discuss your answer.*
3. *Do you "believe" this advertisement? How does the ad differ from a news article on the same subject?*

COMMUTING

by Red Barber

For many years in New York, Red Barber was a broadcaster, first working for the Brooklyn Dodgers and later the New York Yankees. Before he left spring training and headed north in 1939 for his first year as announcer in New York, he talked with Larry MacPhail, who had taken over the Dodger organization.

"I remember asking MacPhail where to live in New York. He went into a lengthy outline. For the first time I heard of Long Island, New Jersey, and Westchester county. The more he talked of various places the more confused I got. Finally he said, 'No matter where you live in New York you're going to have to commute.' This word 'commute' was brand new, but I soon learned what it meant."

Red Barber had a choice between commuting by car or train, and each had its advantages. His house in Scarsdale was two miles from the train station. "The train took about 45 minutes, and the subway from Grand Central to the ball park [Ebbet's Field], 30 minutes. Coming back it was a matter of sheer luck when you got a train. Sometimes on weekends when the train schedule was light I might wait 45 to 60 minutes. Naturally I drove for night games. The trains late at night were few and far between. Driving meant I didn't have to wait and could listen to music and news on the car radio. On the other hand, taking the train into the city gave me a chance to do statistical chores and read the papers, and coming home I could nap.

"However, everything comes at a price, and when you commute you pay for it. How much of my life was spent commuting? How much of my strength went into it? I learned thoroughly in 28 years what MacPhail meant when he warned me that no matter where I live in or around New York I'd have to commute."

QUESTIONS: IDEAS TO THINK ABOUT

1. *Do you commute to work? Describe your trip.*
2. *What choices do you have as to how you get to work? What are the advantages of each?*

"OLD STYLE RAILROADING ON THE GLENDALE LINE"

Despite the everyday problems of the commuter railroads—delays, broken equipment, angry passengers—there remains a bit of the old railroading spirit on the Long Island Railroad. The line built from Jamaica through Glendale and along Newtown Creek to Lond Island City (the Montauk Division) is still in operation, though today, there is only one train a day westbound and two eastbound.

The following article from the Long Island Press in 1936 explains that the ride offered "all the fun of a long railroad trip in 20 minutes," implying that it differed from everyday commuting.

The photo of the Long Island City terminus taken during its heyday in 1895, fifteen years before the opening of the Pennsylvania tubes shows how busy the station was and why it made Long Island City a transportation hub. "Progress," however, left this bit of an earlier era behind.

TO LONG ISLAND CITY		
	AM	
LV Jamaica	8 13	
AR Richmond Hill	8 19	
AR Glendale	8 24	
AR Fresh Pond	8 28	
AR Haberman	8 31	
AR Penny Bridge	8 34	
AR Long Island City	8 41	
	AM	

FROM LONG ISLAND CITY		
	PM	PM
LV Long Island City	4 52	5 22
LV Penny Bridge	4 59	5 28
LV Haberman	5 01	5 30
LV Fresh Pond	5 06	5 35
LV Glendale		5 38
LV Richmond Hill		5 42
AR Jamaica	5 16	5 50
	PM	PM

The photo of the Long Island City terminus taken during its heyday in 1895, fifteen years before the opening of the Pennsylvania tubes shows how busy the station was and why it made Long Island City a transportation hub. "Progress," however, left this bit of an earlier era behind.

QUESTIONS: IDEAS TO THINK ABOUT

1. *Have you ever taken a long railroad trip? When? Describe your journey.*
2. *What are the advantages of rail travel as opposed to flying or driving between cities? The disadvantages?*

Locomotive on the Glendale Spur of the Long Island Railroad

Old Style Railroading on Glendale Line

Steam Locomotives Puff Along Pleasant Route

By SARAH WILFORD

Glendale people who set their clocks in the morning by the whistle of Long Island railroad locomotives on the Jamaica-Glendale-Long Island City line wonder how it happens that steam-drawn passenger trains are still being run so far inside the city limits.

Sarah Wilford

They want to know how many people use this service, and why the railroad keeps it up, and they feel—being rather sentimental people—that, in a city of seven million, in a chromium-plated, stream-lined age, a nice old puffing locomotive is a friendly thing to have around the neighborhood.

The questions are easy to answer. The Long Island railroad runs steam-drawn commuters trains between Jamaica and Long Island City via Richmond Hill, Glendale, Fresh Pond, Penny Bridge and Hunterspoint avenue, because the line has never been electrified. That's simple.

Only 6 Passenger Trains

The railroad keeps on running the passenger service because an average of 1323 people want to use it each weekday morning, and 222 people each weekday afternoon. There are only six passenger trains on the time table. Four leave Jamaica between 7 and 9 a. m. TWO come "home at eventide."

Wonder what happens to the extra trains? They must be piling up at Long Island City. It's something like the puzzle of the extra horse the Department of Plant and Structures discovered when it was making a survey of traffic over the Queensboro Bridge.

Ten horses come over the bridge to Queens each day, and only nine go back to Manhattan.

It begins to look serious when you consider the problem of the extra horses plus the extra trains and add the totals up for a year or two.

You can understand about the extra passengers. They go home on the elevated.

The railroad looks at the line with some of the same sentimental feelings that the residents of Glendale have for the locomotives. It was once the most important part of the most important branch of the whole Long Island system—the Montauk branch.

Between 1860 and 1910 when the East River tunnels were opened the finest trains on Long Island ran on the tracks now used by the six passenger trains and a few freight trains. The passengers were unloaded at the Long Island City depot and crossed to Manhattan on the 34th street ferry.

Tunnels Brought New Route

When the tunnels were built a new route to Jamaica was laid out north of the old one, and the new route, the main line through Woodside that we now use, was electrified immediately.

Business on the old line dropped down and down but never completely stopped although it's just about as easy and in fact a little quicker to get to Long Island City by taking the electric train to Woodside and then transferring to the subway.

The train that has the most passengers is the one that leaves Jamaica at 7:22 a. m. and arrives at Long Island City at 7:42. It carries 418 people on an average day, and is the only one that is really worth running (from the railroad's point of view). The passengers are mostly people who live out on the Island and work in Long Island City factories and warehouses. They all have to be at work early and by 8:35, the time that the last train leaves Jamaica, the number of commuters bound for Long Island City drops to 122.

It was the 8:35 that we took the other day when we made the trip, and a swell train it is.

Everybody seemed to know everybody else. There was plenty of room. Plenty of air. Plenty of sunshine.

All the fun of a long railroad trip in 20 minutes.

Threads Past Library

You roll out of the Jamaica station and into Richmond Hill, past the back of the yellow brick library, looking down on pleasant tree-shaded streets and comfortable homes. Then you shoot through Forest Park—which might just as well be real woods a hundred miles from New York. The plume of smoke from the engine bounces through the bare branches and there are still patches of grey snow on the north side of the wooded banks.

From Forest Park it's only a moment to Glendale, looking all scrubbed and fresh in the morning sunshine.

After Glendale come the cemeteries, miles of them, and after that the factories, first scrubby looking ones, then the huge foundries and rows of grey oil tanks.

We saw those oil tanks on fire once and we'll never again be able to pass them without a slight sinking in the pit of the stomach. It was pretty awe inspiring.

Even before you reach the big factories you catch a glimpse of Newtown Creek which plays hide and seek with the railroad tracks for the last mile or two of the run.

Barges Nearby

Several barges are drawn up beside the tracks in one spot and as we passed we saw a woman in a white apron washing the windows of her floating bungalow. She looked rather angrily busy and we imagine she will be very glad to leave that smoky place and be off again on the up-river journey.

Our trip ends (nowhere) in the railroad yards, a good block or two from the boarded-up old station. From then on it's every man for himself.

You "watch out for the locomotives," dodge freight trains, trip over railroad ties—and come back to Jamaica some other way.

THE SUBWAY AND ELEVATED

Although trolley lines linked the small villages of Queens and carried people to the ferries which took them to "the city," the borough remained relatively isolated. What really opened the interior of Queens to urban development was the construction of new elevated subway lines which connected interior Queens with Manhattan.

The IRT Second Avenue EL was extended to Astoria in 1917, and joint service with the BMT (the RR train) began in 1923. The IRT Flushing Line (#7 train) began service to 103rd Street in Corona in 1917, and reached Main Street, Flushing, in 1928. BMT service to Metropolitan Avenue in Middle Village (M train) began in 1914, to Jamaica (J train) in 1918, and to Ridgewood along Wyckoff Avenue (LL train) in 1924. The IND line along Queens Boulevard (E and F trains), today the most heavily traveled subway in Queens, was not opened until 1933, with the final stretch to 179th Street in Jamaica completed in 1950.

Communities quickly felt the effects of the new elevated lines. Within a few years farmland sprouted homes, schools, and apartment buildings instead of cabbages and the people of Queens welcomed the changes. The trains cut traveling time to Manhattan in half. After the completion of the Atlantic Avenue tubes in 1904, communities as far away as Jamaica held "subway celebrations," parades in honor of the new link with the city. The subway was a sign of progress.

THE JAMAICA SUBWAY CELEBRATION, 1908. This horse-drawn trolley of the old Brooklyn and Jamaica Rail Road came out of retirement for the 1904 Jamaica "Subway Celebration," an event that marked the opening of the subway tunnel between lower Manhattan and Atlantic Avenue in Brooklyn. The celebration featured various forms of transportation from the past, present, and future, including this horsecar, a stagecoach, bicycles, automobiles, and even a flying machine.

QUESTIONS: IDEAS TO THINK ABOUT

1. *Why was the opening of the subway an important event for Jamaica? Who took part in the celebration?*
2. *Have you witnessed or taken part in any similar community celebrations or parades?*

OPENING OF THE STEINWAY TUNNEL, MAY 22, 1915. This photograph shows the opening of the Steinway Tunnel: workmen and area residents watch a trolley car emerge after making the first run between Grand Central Station and Vernon Avenue. This was the first car to run under the East River to Queens.

ELEVATED CONSTRUCTION ON ROOSEVELT AVENUE, 1914.

BUILDING THE ELEVATED, 1915

The IRT Flushing line, which runs along Roosevelt Avenue, caused tremendous changes in Queens. Most of the land where the line was built was nothing but farms, with no large communities along the route. This photograph of the EL under construction demonstrates the startling contrast between the elevated railroad and the dirt road, with large trees growing directly underneath the steel girders.

QUESTIONS: IDEAS TO THINK ABOUT

1. *It seems that the El was built in the middle of nowhere. Why? Who was this line meant to serve?*
2. *How would we react today if a transit line was built where few people could use it?*

"SIXTH AVENUE ELEVATED AFTER MIDNIGHT," EVERETT SHINN, 1899.

"The apparation of these faces in the crowd,
Petals on a wet black bough."
"In a Station of the Metro," by Ezra Pound

Once the novelty of the subway wore off, it settled into its function of carrying people to work, or home from the city after a long night. The subway made possible the separation of workplace and residence and increased the labor pool for many businesses. An artist or poet observing a station or a car filled with lonely, tired people becomes aware of the darker, perhaps sordid, aspects of riding the trains, the worn faces and grimy windows.

QUESTIONS: IDEAS TO THINK ABOUT

1. *How do the painting and the words of the poem create a particular mood?*
2. *What is Shinn's or Pound's point about the subway? Have you ever experienced similar feelings?*
3. *The Transit Authority has considered closing the subways between midnight and 5 a.m. What is your opinion of this idea? Why?*
4. *Riding the subway in New York is often depressing. Why? (Grafitti, filth, delays, other passengers, etc.)*

THE AUTOMOBILE

The change from the horse and buggy era to the automobile age did not happen all at once. For many years it was common to see cars and horses sharing the road. As more Americans could afford a car or a truck, wagons became obsolete. Still, in some places the horse remained the preferred mode of transportation.

Although the automobile had been available since the 1890's it was not until Henry Ford introduced the Model T in 1908 that the automobile became affordable to the majority of people for the first time. By 1916, Model T sales had jumped from under 6,000 per year to nearly 600,000, while the cost dropped from $950 to $360.

The automobile brought about changes in the landscape as well as life style. It created a demand for new and improved roads and services, but the technology for the new age was not immediately in place. The impulse to get behind the wheel and race headlong into the future was tempered by, among other things, dirt roads that naturally turned to mud in the rain.

The documents in this section examine the automobile from various perspectives: how cars function as status symbols; the experience of making, selling and buying cars; the experience of driving through a changing concrete landscape. The song opening this section, "In My Merry Oldsmobile," demonstrates that cars also captured the popular imagination and have continued to be a powerful image in popular culture.

Pictured here is 13 year old Catharine Rehkamp and her brother John behind the wheel.

THE REHKEMP FAMILY CAR, 1913

The Rehkamp family owned a large and successful saloon in Ridgewood, and they also had one of the first cars. Pictured here is 13 year old Catharine Rehkamp, sitting in the back seat with a bow in her hair, and her brother John behind the wheel. Earlier that day she was released from the hospital where she had her appendix removed. They are parked in front of their saloon; a sign in the window advertises, "pure home made wines, 5¢ 1/2 schoppen." The automobile expanded the family's business as John often used the car to deliver home made wine and bottled beer to their customers in the neighborhood. It is not much different from the way Mrs. Geipel used her horse and buggy.

Whenever new forms of transportation become available to people, life changes considerably, often in unanticipated ways. The car changed business and family life by increasing the markets for family businesses and by introducing the "car culture," emphasizing style, speed and status.

QUESTIONS: IDEAS TO THINK ABOUT

1. *How did the Rehkamps use their car for business?*
2. *How did the car reflect their standing in the Ridgewood community?*

THREE WOMEN IN PHOTO STUDIO, 1918

These women understand the status involved in owning a car. They posed behind the wheel of a cardboard automobile in a photo studio on Jamaica Avenue. A car represented everything these women wanted for their family; like home ownership, a car meant they had "arrived." In reality, however, they worked hard and struggled not to buy a car, but to send their brother to college, and later to medical school. That decision determining how the family spent their money represented the difference between looking successful and real "moving up" through achievement.

QUESTIONS: IDEAS TO THINK ABOUT

1. *Why is owning a car important?*
2. *Do you remember your first car? What did you give up in order to afford a car? How did you finance your car?*

FRED SEYMOUR AND HIS CAR, 1939.

THE AUTOMOBILE INDUSTRY IN LONG ISLAND CITY

After the Queensboro Bridge opened in 1909, industrialists took advantage of the available land in Long Island City, the perfect site for new factories. In addition there was a growing workforce and now mass transit to carry the workers to the factories.

Long Island City was once one of the centers of automobile manufacturing in the United States. According to the 1915 yearbook of the Queens Borough Chamber of Commerce, there were 17 automobile and truck factories in Queens, employing almost 1,600 workers. Among them were Ford, Packard, Brewster, Pierce Arrow, and White Trucks. Other manufacturers supplied the car factories with wire wheels, tires, window glass, metal fixtures, lubricants, and horns.

HARROLDS MOTOR CAR CO. (PIERCE-ARROW SERVICE PLANT), LONG ISLAND CITY, c. 1909.

This industrial growth meant several things to residents of the area. First it created new jobs. Secondly, it crowded out the old neighborhoods, and thirdly it increased air and water pollution. It also brought banks to the area with money for investment in new business.

One resident remembers the effect the industrial development had on his own life. George Steinmayer was born in Long Island City and grew up there. When he returned home from World War I in 1919 the country was in a recession, he could not find steady work: "it was odd jobs; there was a depression. There were no jobs ... very few people had steady jobs, especially war veterans." He took a temporary job in the storeroom at Brewster Automobiles, filling in for two weeks for a vacationing worker. At the end of two weeks the man did not return, so the job was George's. He worked there until 1936, when the factory was closed, one of many companies forced out of business by the Great Depression.

ROLLS-ROYCE PHAETON, 1929.

...I was a materials man. We had a parts department, not just a parts dept., we kept everything needed to build an automobile—metals, fabrics, paints. I worked there from 1920 to 1936. ... They were building Brewster automobiles; they were built complete in that one building, everything from a chassis to a body to a 4 cylinder motor... they made the engines ... everything, the gears, transmissions ... Rolls Royce bought out Brewster in 1925. Brewster was having difficulties, but also they were competing with Rolls Royce. Most people in N.Y. did not want an English body, they wanted a Brewster body. That was a name that was glorious, like Steinway is.

George Steinmayer

G. V. ELECTRIC TRUCKS

Here is a 2-ton G. V. Electric operated by the New York & Queens Electric Light & Power Company. It is one of 19 used by this progressive Central Station, which serves Queens borough business men and Queens borough homes so well.

On another page you will find other data about the world's largest Electric truck plant —in Queens borough. There are over 1,500 G. V. Electric trucks and wagons in operation in Greater New York. How about one or two for your delivery service? Catalogue? Certainly.

GENERAL VEHICLE COMPANY, INC.
General Office and Factory, Long Island City, N. Y.

NEW YORK CHICAGO BOSTON PHILADELPHIA

QUESTIONS: IDEAS TO THINK ABOUT

1. *Why did automobile companies locate in Queens? Why are there no car factories in Queens today?*
2. *Did you or anyone you know work in one of those factories? Describe your experience.*
3. *What does it mean for a community when such large factories are built nearby? What are the benefits? The problems?*

BABBITT

by Sinclair Lewis

The following selection from *Babbitt*, Sinclair Lewis' great novel about American life in the 1920's, shows the attitude of a middle class family toward buying a new car. Each member of the family wants something different in the new car. The 1920's were prosperous years when almost all middle-class families could look forward to owning a car.

George Babbitt starts the discussion at the dinner table, after the excellent "roast of beef, roasted potatoes, and string beans."

BABBITT

"Sort o' thinking about buying a new car. Don't believe we'll get one till next year, but still, we might."

Verona, the older daughter, cried, "Oh, Dad, if you do, why don't you get a sedan? That would be perfectly slick! A closed car is so much more comfy than an open one."

"Well now, I don't know about that. I kind of like an open car. You get more fresh air that way."

"Oh, shoot, that's just because you never tried a sedan. Let's get one. It's got a lot more class," said Ted.

"A closed car docs keep the clothes nicer," from Mrs. Babbit; "you don't get your hair blown all to pieces," from Verona; "It's a lot sportier," from Ted; and from Tinka, the youngest, "Oh, let's have a sedan! Mary Ellen's father has got one." Ted wound up, "Oh, everybody's got a closed car now, except us!"

Babbitt faced them: "I guess you got nothing very terrible to complain about! Anyway, I don't keep a car just to enable you children to look like millionaires! And I like an open car, so you can put the top down on summer evenings and go out for a drive and get some good fresh air. Besides— A closed car costs more money."

"Aw, gee whiz, if the Doppelbraus can afford a closed car, I guess we can!" prodded Ted.

"Humph! I make eight thousand a year to his seven! But I don't blow it all in and waste it and throw it around, the way he does! Don't believe in this business of going and spending a whole lot of money to show off and—"

They went, with ardor and some thoroughness, into the matters of streamline bodies, hill-climbing power, wire wheels, chrome steel, ignition systems, and body colors. It was much more than a study of transportation. It was an aspiration for knightly rank. In the city of Zenith, in the barbarous twentieth century, a family's motor car indicated its social rank as precisely as the grades of the peerage determined the rank of an English family . . . and where Babbitt as a boy had aspired to the presidency, his son Ted aspired to a Packard twin-six and an established position in the motored gentry.

Sinclair Lewis

QUESTIONS: IDEAS TO THINK ABOUT

1. *From this brief passage, what do you know about the Babbitt family?*
2. *How does your family decide on major purchases such as a car or a house?*

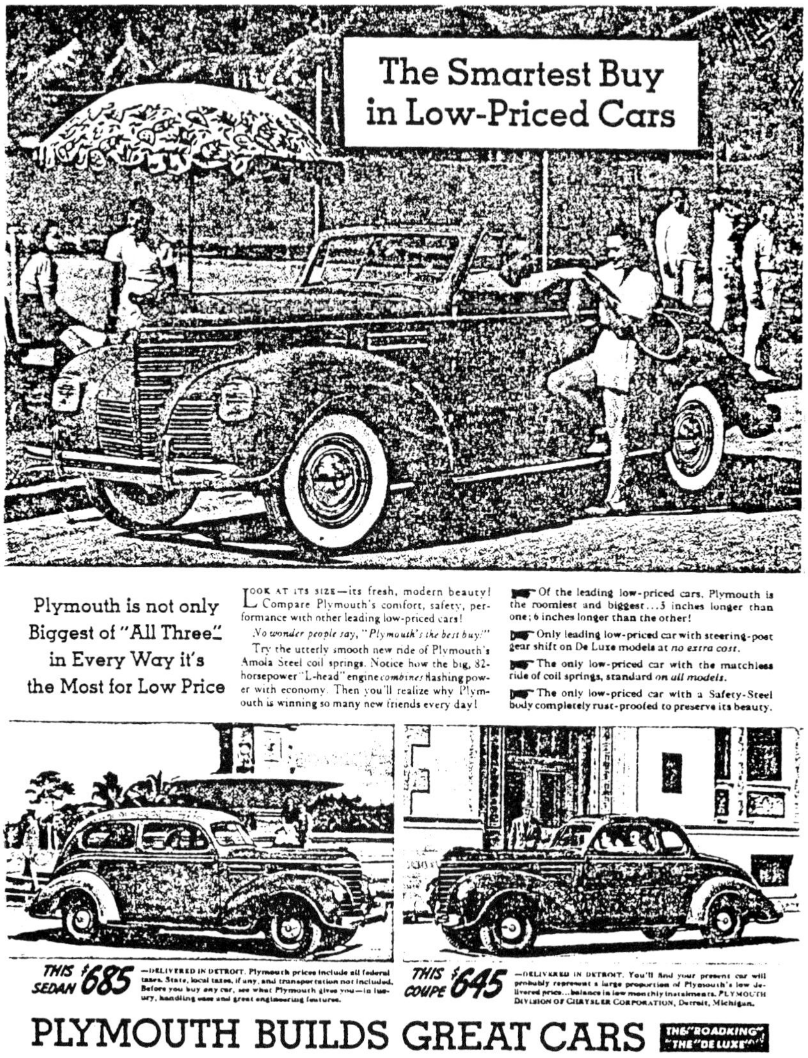

ADVERTISEMENT FROM LIFE MAGAZINE, 1939.

The car had once been nothing but a curiosity, then a status symbol, and finally an accepted part of everyday life. Once America was sold on the idea of the private automobile, advertisements focused on the special features of a particular model or the image of one car as compared with another.

In 1939, the date of this Plymouth ad, America was still in the grip of the Depression, and few working people could afford a new car, even at the "low price" of $645. Yet the ad does not shrink from portraying the "good life" associated with owning a new car.

ADVERTISEMENTS FROM LIFE MAGAZINE, 1941.

Manufacturers made a variety of appeals for people to buy their cars, emphasizing luxury or practicality, sportiness or efficiency, the thrill of driving and the ease of the controls. We can see how these advertisements appealed to the attitudes the Babbitt family expressed over their dinner table twenty years before.

QUESTIONS: IDEAS TO THINK ABOUT

1. *From the Plymouth ad (1939) would you know the country was in a depression? Why or why not?*
2. *What are the selling points in each ad?*
3. *How does each ad create an image of the driver of that car?*
4. *What is the image of women in the Oldsmobile ad? Do you think it is accurate?*
5. *Compare these ads with automobile ads from the present. How are the appeals similar or different?*

NORTHERN BOULVARD (ROUTE 25A) IN *THE GREAT GATSBY, 1925*

The Great Gatsby, F. Scott Fitzgerald's well-known novel, is not about Queens, but it does offer a glimpse of what it was like to motor through the borough before parkways stretched out to Long Island. To reach West Egg (Great Neck) from New York, Jay Gatsby travelled along Northern Boulevard, route 25A, from the Queensboro Bridge, passing the factories of Long Island City, the Corona dumps where in 1939 the World's Fair would rise, and through downtown Flushing. The route had been used since 1860, the first direct road between the 34th Street Ferry and Flushing.

Fitzgerald himself moved to Great Neck in 1922. To reach New York he either took the Long Island Railroad or drove along Northern Boulevard. In *The Great Gatsby*, Fitzgerald heightened his own impressions of the journey. A functional bridge became a gateway to the world's fortunes; a landfill became a wasteland. He describes the train ride through the Sunnyside railroad yards:

The next day was broiling, almost the last, certainly the warmest, of the summer. As my train emerged from the tunnel into sunlight, only the hot whistles of the National Biscuit Company broke the simmering hush at noon. The straw seats of the car hovered on the edge of combustion.

We do get an impression however of what the small towns of Queens looked like before the full impact of urbanization was felt. Through Fitzgerald's descriptions we can recognize features of Queens' landscape in the 1920's.

About half way between West Egg and New York the motor road hastily joins the railroad and runs beside it for a quarter of a mile, so as to shrink away from a certain desolate area of land. This is a valley of ashes—a fantastic farm where ashes grow like wheat into ridges and hills and grotesque gardens; where ashes take the forms of houses and chimneys and rising smoke and, finally, with a transcendent effort, of men who move dimly and already crumbling through the powdery air. Occasionally a line of gray cars crawls along an invisible track, gives out a ghastly creak, and comes to rest, and immediately the ash-gray men swarm up with leaden spades and stir up an impenetrable cloud, which screens their obscure operations from your sight.

The valley of ashes is bounded on one side by a small foul river, and, when the drawbridge is up to let barges through, the passengers on waiting trains can stare at the dismal scene for as long as half an hour. There is always a halt there of at least a minute, and it was because of this that I first met Tom Buchanan's mistress.

Over the great bridge, with the sunlight through the girders making a constant flicker upon the moving cars, with the city rising up across the river in white heaps and sugar lumps all built with a wish out of non-olfactory money. The city seen from the Queensboro Bridge is always the city seen for the first time, in its first wild promise of all the mystery and the beauty in the world.

F. Scott Fitzgerald

NORTHERN BOULEVARD IN JACKSON HEIGHTS, 1925.

QUESTIONS: IDEAS TO THINK ABOUT

1. *How does Fitzgerald use changes in transportation to develop his descriptions?*
2. *Read the whole novel and discuss how the author sees the role of the automobile in people's lives.*

HIGHWAYS AND TRAFFIC JAMS

On weekends during the prosperous 1920's the outdated roads were jammed bumper to bumper with families trying to go for a ride in the country. When people finally arrived at a destination away from the city, they found that almost all of the countryside they came to see was "private property."

Robert Moses, a city planner and Parks Commissioner in the 1930's, worked to change that situation, building parkways out to Long Island, direct highways that carried people to parks like Jones Beach, which he also built. For Robert Moses, the automobile was mass transit of the future. The irony however, was that no sooner did a new parkway open than it was jammed with more cars than it was meant to support.

GRAND CENTRAL PARKWAY AND NORTHERN BOULEVARD, 1939.

QUESTIONS: IDEAS TO THINK ABOUT

1. *Describe your feelings when you are caught in a traffic jam.*
2. *Do you remember when a highway was built? Was it an improvement? What kinds of roads did the highway replace?*
3. *Who was Robert Moses? What was his effect on transportation in New York City? How did his projects affect your neighborhood?*

Every advance in technology usually brings other problems. Karl Shapiro, a modern poet, reminds us of one of the negative aspects of the automobile age, the loss of life from accidents.

"AUTO WRECK"

Its quick soft silver bell beating, beating
and down the dark one ruby flare
Pulsing out red light like an artery,
the ambulance at top speed floating down
Past beacons and illuminated clocks
Wings in a heavy curve, dips down,
And brakes speed, entering the crowd.
The doors leap open, emptying lights;
Stretchers are laid out, the mangled lifted
And stowed into the little hospital,
Then the bell, breaking the hush, tolls once,
And the ambulance with its terrible cargo
Rocking, slightly rocking, moves away,
As the doors, and afterthought, are closed.

We are deranged, walking among the cops
Who sweep glass and are large and composed.
One is still making notes under the light.
One with a bucket douches ponds of blood
into the street and gutter.
One hangs lanterns on the wrecks that cling,
Empty husks of locusts, to iron poles.

Our throats were tight as tourniquets,
Our feet were bound with splints, but now,
Like convalescents intimate and gauche,
We speak through sickly smiles and warn
With the stubborn saw of common sense,
The grim joke and the banal resolution.
The traffic moves around with care,
But we remain, touching a wound
That opens to our richest horror,
Already old, the question who shall die?
Becomes unspoken who is innocent?

For death in war is done by hands;
Suicide has cause and stillbirth, logic;
And cancer, simple as a flower, blooms.
But this invites the occult mind,
Cancels our physics with a sneer,
And spatters all we knew of denouement
Across the expedient and wicked stones.

Karl Shapiro

QUESTIONS: IDEAS TO THINK ABOUT

1. *How does this poem convey the experience of an accident? Is the poet successful?*
2. *Have you ever been in an accident? Describe what happened. What problems resulted?*
3. *What can be done to lessen the severity of car accidents? Do you wear seatbelts? What is the government's role in questions of auto safety?*

THE AIRPLANE

Flying has always appeared to individuals wanting a thrill or a challenge. Over the years, ordinary individuals like John Abramopaulis and heroes like Charles Lindbergh or Amelia Earhart have responded to the challenge and have added their contributions to aviation history.

Flying is perhaps the most exciting way of moving that we can experience today. The spread of airlines and the increase in air traffic affects our lives significantly. People who live in Queens complain about the noise of the airplanes; they petition the airlines to change the flight patterns. Most communities resist having an airport built nearby because of the traffic and the pollution. Yet, millions take advantage of the speed offered by the airplane and of the way it has brought the world closer together. At the same time, flight can bring disaster; it gives countries the capacity to bomb each other and to make war on a horrifying scale.

The Wright Brothers built their airplane in their bicycle shop in 1903, but not until World War I were aircraft designs improved so that after 1920 air travel became a reality. The greatest advances in machine design usually came in times of war or in exploring new frontiers like the flight to the moon.

As with every other advance in technology, the field of air travel has created whole new industries and jobs, and has changed business patterns. People now commute to work, from New York to Chicago by airplane. Business people can fly from New York to Boston or Washington and do a day's work and can be home for dinner.

QUEENS AIRPORTS

In November 1930, the publication of the Queensboro Chamber of Commerce boasted that:

Queens, ... with its almost daily air events of local, national and international importance, is justly proud of the lead it has taken in aviation development including the construction of complete air terminals with full facilities for handling every type of passenger, mail and commercial service.

The men who have put their time and energies and money into these vast projects have done so, not only as pioneers, but as sound, level-headed business men who have seen how quickly air service is coming on the nation; how important suitable airports are to this service and how important they are to New York City, if it is to keep its lead in this new and growing enterprise as it has in others.

The principal airfields in Queens at that time were Holmes Airport, where John Abramopaulos took his first flight, Curtis Field at North Beach, and Flushing Airport, small fields unable to handle large commercial aircraft. Brooklyn's Floyd Bennet Field opened as the city's first municipal airport in 1931. It was a long ride, however, from the foot of Flatbush Avenue where the airport was located, to Manhattan. Soon plans were being made for a second municipal airport to be built in Queens.

Building LaGuardia Airport during the Depression, when one out of four Americans were out of work, was an enormous undertaking for the financially strapped city. Through the Works Progress Administration (WPA), a program initiated by the Roosevelt Administration to help local municipalities, the city was able to employ thousands of men and women.

PLANS FOR LAGUARDIA AIRPORT. Built on the site of North Beach, the turn of the century amusement area closed by Prohibition, LaGuardia Airport opened in 1939. In the aerial view, between the airport and the distant skyline lies the village of Astoria. Note the Queensboro and Triboro Bridges and the elevated subway line.

QUESTIONS: IDEAS TO THINK ABOUT

1. *Why were several airports located in Queens?*
2. *How does having an airport nearby affect a community?*
3. *Why does the city put part of its resources into making air travel easier? Should the "Train to the Plane" continue, even though it loses money?*

CAPTAIN BEACHLEY'S AIRSHIP, 1908

The Jamaica "Subway Celebration" was in fact a celebration of all transportation, past, present and future. Everything from a stagecoach to a horseless carriage took part in a parade down Jamaica Avenue. The parade even included a look into the future: a demonstration of Captain Beachley's Airship. This flimsy, human-propelled contraption actually did get off the ground, to the delight of the crowd.

QUESTIONS: IDEAS TO THINK ABOUT

1. *How did Beachley's airship work?*
2. *Have you ever gone hang-gliding? Describe the experience.*

JOHN ABRAMOPAULOS: THE THRILL OF FLYING

As a boy, in the late 1920's, John Abramopaulos often walked to Holmes Airfield, only two miles from his home in Astoria. The grass field ran from Northern Boulevard to Astoria Boulevard, just an oblong plot of open ground. He describes his first ride in this way:

I believe at that time the Empire State Building was just built ... '31 or '32. They were selling rides ... over the Empire State Building and back from Holmes Airport. I think it was $2.50 ... a lot of money in those days. I managed to save. I used to work after school; I guess I must've been 14, 13. I worked in a fruit store. I used to hate it but they paid me. ... They put four of us in an old 'Standard' ... a good solid biplane. ... They flew us over to the Empire State Building and we flew around the top of it and just before we landed the pilot pulled the plane up and down, gave us sort of a thrill. ... We got a big kick out of it.

Later Mr. Abramopaulos took flying lessons out at Flushing Airport. "In those days," he recalls, "it was a series of puddles so we had to land between the puddles. I flew with Speed Hanslick for a while." He continued his lessons, paying for one at a time, out at Roosevelt Field on Long Island, where Charles Lindbergh had taken off for his historic solo flight across the Atlantic. In his own way, John became a daring flyer. "I've done everything up there," he says, "aerobatics ... I like to flip an airplane around. ... They used to tell me when I used to fly I never used to stand still. I was always doing something. I used to enjoy doing spins."

Since his retirement, Mr. Abramopaulos has been working on a vehicle that can be used on the road as well as in the air:

> "One thing they never tell you. After you land at a field, what do you do? You can't get home. ... You've got 10 or 20 thousand dollars tied up in an airplane that's not doing anything anymore for you. That's one of the reasons, I believe, that a lot of people don't fly. Where they can afford one vehicle they're forced to have two. The vehicle they get is the one that's most useful to them, and that turns out to be a car. ... Unless the big companies develop a vehicle that you can take home with you, land at a field and drive it home. ... I've worked on that myself. ... Why haven't the companies done something like that, and really make the airplane a functional unit, instead of an expensive hobby; that's all it is today."

The problems Mr. Abramopaulos talks about are the same ones facing the airlines today. The planes can carry 350 people, but they cannot speed up the loading and unloading of people and they cannot improve the traffic on the ground.

John Abramopaulos (on the left) with his plane, a 1929 Fleet, and his friend. This was John's second plane, which he rebuilt at Flushing Airport. He had taken flying lessons at Holmes Airport, a grass field located at Northern Boulevard and 70th Street and stretching to Astoria Boulevard. After getting his pilot's license, John flew out of Flushing Airport, Roosevelt Field, and airfields in New Jersey and upstate New York.

QUESTIONS: IDEAS TO THINK ABOUT

1. *Have you ever flown in a private airplane? How is the experience different from flying in a commercial airlines?*
2. *Describe your first flight.*
3. *Have you ever taken flying lessons?*
4. *What do you think of Mr. Abramopaulos' idea for "roadable autogyro"? Would you buy one?*

HEROES OF AVIATION

The exploits of all the heroes of aviation, from Charles A. Lindbergh to Yuri Gagarin, have set off similar feelings of pride and wonder. But what did it mean to the millions on both sides of the Atlantic who cheered Lindbergh's pioneering solo flight? What was the effect on their personal lives of the exploits of Amelia Earhart, the first woman to fly across the Atlantic? What exactly were they cheering?

Charles Lindbergh, the first man to fly alone across the Atlantic in 1927 writes about his thoughts and feelings as he takes off from Roosevelt Field in Mineola:

THE SPIRIT OF ST. LOUIS

Thirty Revolutions Low! The engine's vibrating roar throbs back through the fuselage and drums heavily on taut fabric skin. I close the throttle and look out at tense faces beside my plane. Life and death lies mirrored in them—rigid, silent, waiting for my word.

Thirty revolutions low—a soft runway, a tail wind, and overload. I glance down at the wheels. They press deeply, tires bulging, into the wet, sandy clay.

My cockpit quivers with the engine's tenseness. But the Spirit of St. Louis isn't vibrant with power as it's always been before. I'm conscious of the great weight pressing tires into ground, of the fragility of wings, of the fullness of oversize tanks of fuel. There is in my plane this morning, more of earth and less of air than I've ever felt before.

Plane ready; engine ready; earth-inductor compass set on course. The long, narrow runway stretches out ahead. Over the telephone wires at its end lies the Atlantic Ocean; and beyond that, mythical as the rainbow's pot of gold, Europe and Paris. This is the moment I've planned for, day and night, all these months past. The decision is mine. No other man can take that responsibility. The mechanics, the engineers, the blue-uniformed police officers standing there behind the wing, everyone has done his part. Now, it's up to me.

Charles A. Lindbergh

QUESTIONS: IDEAS TO THINK ABOUT

1. *Why is the public inspired by the exploits of Lindbergh, Earhart, or the astronauts?*
2. *How do your memories of the events compare with the newspaper accounts?*

BIBLIOGRAPHY

WORK HISTORY

Anderson, Jervis, *A. Philip Randolph: A Biographical Portrait,* N.Y.: Harcourt, Brace, Jovanovich, 1972.

Anthony, Katherine I., *Mothers Who Must Earn,* N.Y.: Russell Sage Foundation, 1914.

Blumberg, Barbara M., *The New Deal and the Unemployed: the View from New York City,* Lewisburg: Bucknell University Press, 1979.

Bureau of Women in Industry, *Hours and Earnings of Women Employed in Power Laundries in New York State,* Special Bulletin No. 153, 1927.

Campbell, Helen, *Prisoners of Poverty: Women Wage Earners, Their Trade and Their Lives,* Boston: Roberts Brothers, 1887. (Reprint: Westport: Greenwood Press, 1970.)

Capeci, Dominic J., Jr., *The Harlem Riot of 1943,* Philadelphia: Temple University Press, 1977.

Caro, Robert, *The Power Broker: Robert Moses and the Fall of New York,* N.Y.: Vintage, 1975.

Chapin, Robert C., *The Standard of Living Among Workingmen's Families in New York City,* N.Y.: Russell Sage Foundation, 1909.

Clark, Sue and Edith Wyatt, *Making Both Ends Meet: The Income and Outlay of New York Working Girls,* N.Y.: Macmillan, 1911.

Cohen, Mini, "Italian American Women in New York City, 1900-1950: work and school," in *Class, Sex, and the Woman Worker,* ed. by Milton Cantor and Bruce Laurie, Westport: Greenwood, 1977.

Cott, Nancy F. and Elizabeth H. Pleck. *A Heritage of Her Own.* New York: A Touchstone Book, 1979.

Dublin, Thomas. *Women at Work.* New York: Columbia University Press, 1979.

Dubofsky, Melvin, *When Workers Organize: New York City in the Progressive Era,* Amherst: University of Massachusetts, 1968.

Filley, Jane and Therese Mitchell, *Consider the Laundry Workers,* N.Y.: New York League of Women Shoppers, June 1937.

Genovese, Eugene D. *Roll, Jordon, Roll: The World The Slaves Made.* New York: Vintage Books, 1976.

Guinier, Ewart. "Impact of unionization on blacks," in *Unionization of Municipal Employees,* ed. by Robert Connery and William Farr, N.Y. Academy of Political Science, 1970.

Gutman, Herbert. *The Black Family in Slavery and Freedom, 1750-1925.* New York: Vintage Books, 1977.

Gutman, Herbert. *Work, Culture and Society in Industrializing America.* New York: Vintage Books, 1977.

Hall, Max, ed., *Made in New York: Case Studies in Metropolitan Manufacturing,* Cambridge: Harvard University Press, 1959.

Harris, Louis and Nelle Swartz, *The Cost of Clean Clothes in Terms of Health: A Study of Laundries and Laundry Workers in New York City,* N.Y.: Department of Health, 1917.

Haynes, George E., *The Negro at Work in New York City: a Study in Economic Progress,* N.Y.: Columbia University Press, 1912.

Howe, Louise Kapp, *Pink Collar Workers: Inside the World of Women's Work,* New York, Avon Books, 1977.

Huggins, Nathan Ivan. *Black Odyssey: The Afro-American Ordeal in Slavery.* New York: Vintage Books, 1977.

Kwong, Peter, *Chinatown New York: Labor and Politics, 1930-1950,* N.Y.: Monthly Review Press, 1979.

Lerner, Gerda, ed. *Black Women in White America: A Documentary History.* New York: Vintage Books, 1973.

More, Louis B., *Wage Earners' Budgets: A Study of Standards and Costs of Living in New York City,* N.Y.: Holt, 1907.

Nestel, Louis, *Labor Relations in the Laundry Industry in Greater New York,* N.Y.: Claridge, 1950.

New York State Department of Labor, *Working Conditions in New York City Steam Laundries,* Bulletin No. 50, March 1913.

Newfield, Jack and Paul DuBrul, *The Abuse of Power: the Permanent Government and the Fall of New York,* N.Y.: Viking, 1977.

Ovington, Mary, *Half a Man: the Status of the Negro in New York,* N.Y.: Longmans, 1911.

Packard, Esther, *A Study of Living Conditions of Self-Supporting Women in New York City,* N.Y.: YMCA, 1915.

Ringenbach, Paul Thomas, *Tramps and Reformers, 1873-1916: The Discovery of Unemployment in New York,* Westport: Greenwood Press, 1973.

Robin, Lester, *The Negro in the Longshore Industry,* Racial Policies of American Industry, Report 29, Philadelphia: University of Philadelphia Press, 1974.

Scott, Joan Wallach. *The Glassworkers of Carmaux.* Cambridge Massachusetts: Howard University Press, 1974.

Spero, Sterling and Abram Harris, "The Negro Longshoreman, 1870-1938," in *Black Workers and Organized Labor,* ed. by Bailey M. Rudwick, Belmont: Wadsworth, 1971.

Terkel, Studs. *Working: People talk about what they do all day and how they feel about what they do.* New York: Pantheon Books, 1974.

Wertheimer, Barbara. *We Were There: The Story of Working Women in America.* New York: Pantheon Books, 1977.

LEISURE HISTORY

Balio, Tino, ed. *The American Film Industry.* Madison: University of Wisconsin Press, 1976.

Baran, Paul and Paul Sweezy. *Monopoly Capital.* New York: Monthly Review Press, 1966.

Baron, Stanley. *Brewed in America: A History of Beer and Ale in the United States.* Boston: Little, Brown, 1962.

Bengstsson, Arvid, ed. *Adventure Playgrounds.* New York: Praeger, 1972.

Blocker, Jack S. *Retreat from Reform: The Prohibition Movement in the United States, 1890-1913.* Westport, Conn.: Greenwood Press, 1976.

Clark, Norman H. *Deliver Us From Evil: An Interpretation of American Prohibition.* New York: W.W. Norton, 1976.

Cochran, Thomas. *The Pabst Brewing Company: The History of an American Business.* New York: New York University Press, 1948.

DiMeglio, John E. *Vaudeville, U.S.A.* Bowling Green, Ohio: Bowling Green University Popular Press, 1973.

Dulles, Foster Rhea. *America Learns to Play—A History of Popular Recreation.* New York: Appleton-Century, 1940.

Ewen, Stuart. *Captains of Consciousness: Advertising and the Social Roots of the Consumer Culture.* New York: McGraw Hill, 1976.

Furnas, J.C. *The Life and Times of the Late Demon Rum.* New York: G.P. Putnam's Sons, 1965.

Gilbert, Douglas. *American Vaudeville: Its Life and Times.* New York: Dover Press, 1940.

Griffin, Clifford S. *Their Brothers' Keepers: Moral Stewardship in the United States, 1800-1865.* New Brunswick, N.J.: Rutgers University Press, 1960.

Grimsted, David. *Melodrama Unveiled: American Theater and Culture, 1800-1860.* Chicago: University of Chicago Press, 1968.

Gulfield, Joseph. *Symbolic Crusade: Status Politics and the American Temperance Movement.* Urbana, Ill.: University of Illinois Press, 1963.

Jacobs, Lewis. *The Rise of the American Film: A Critical History.* 1939; rpt. New York: Teachers College Press, 1968.

Johnston, Alexander. *Ten—And Out!—The Complete Story of the Prize Ring in America.* New York: Ives Washburn, 1927.

Jowett, Garth. *Film: The Democratic Art.* Boston: Little, Brown, 1976.

Krout, John Allen. *The Origins of Prohibition.* New York: Alfred A. Knopf, 1925.

Lahue, Kalton C. *Continued Next Week: A History of the Moving Picture Serial.* Norman: University of Oklahoma Press, 1964.

Laurie, Joe, Jr. *Vaudeville: From Honky Tonks to the Palace.* New York: Henry Holt, 1953.

Marrus, Michael, ed. *The Emergence of Leisure.* New York: Harper and Row, 1974.

McLean, Albert. *American Vaudeville as Ritual.* Lexington: University of Kentucky Press, 1965.

Rainwater, Clarence. *The Play Movement in the United States: The Study of Community Recreation.* Chicago: University of Chicago Press, 1922.

Robinson, David. *Hollywood in the Twenties.* New York: A.S. Barnes, 1968.

Sklar, Robert. *Movie-Made America.* New York: Random House, 1975.

Wagenknecht, Edward. *The Movies in the Age of Innocence.* New York: Ballantine Books, 1971.

TRANSPORTATION HISTORY

Cheape, Charles W. *Moving the Masses.* Boston: Harvard University Press, 1980.
Condit, Carl W. *The Port of New York. A History of the Rail and Terminal Pennsylvania Station.* Chicago: University of Chicago Press, 1980.
Cunningham, Joseph and DeHart, Lemond. *A History of the New York Subway System.* New York, 1976. 3 Vols.
Electric Railroaders Association. *Newsletter* and *Photo Files.* "Existing and Projected Trolley Lines Around New York." *Harper's Weekly,* Aug. 1, 1896.
Fenelly, Richard A. *People Movers on the New York City Transit System.* NYCTA, 1970.
Fifth Ave. Coach Co. *Motor Bus Relief for New York City's Transit Needs.* New York, 1917.
Fishler, Stan. *Uptown Downtown: A Trip Through Time on New York Subways.* New York: Hawthorne Books, 1976.
Haskell, Gilbert. *The Subway and Tunnels of New York.* New York: Wiley and Sons, 1912.
"How People Come and Go in New York." *Harpers Weekly,* Feb. 26, 1898.
Kahn, Alan Paul. *The Tracks of New York 1907: The Metropolitan Street Railway.* New York: Electric Railroaders Association, 1973.
Kramer, Frederic A. *Across New York By Trolley.* New York: Quadrant Press, Inc., 1975.
"Rapid Elevated Lines in New York City." *New York Historical Society Quarterly,* vol. 18, no. 4, Jan. 1935, pp. 54-82. vol. 19, no. 1, Apr. 1935, pp. 3-17.
Reeves, William Fullerton. *The First Elevated Railroads in Manhattan and the Bronx of the City of New York.* New York: New York Historical Society, 1936.
Spengler, Edwin H. *Land Values in New York in Relation to Transit Facilities.* New York: Columbia University Press, 1930.
Walker, James B. *Fifty Years of Rapid Transit, 1864-1917.* New York: Arno Press, 1970.

QUEENS HISTORY

A History of Long Island City. Garden City: privately printed, 1982.
A History of Queens Village. Uniondale: Salisbury Printer, 1964.
Armbruster, Eugene. *Pictorial Works: Queens, Newtown, Brooklyn.* Developed during the 1940s.
Byers, Pam. *Small Town in the Big City.* A History of Sunnyside-Woodside. Published, 1976.
Caro, Robert A. *The Power Broker: Robert Moses and the Fall of New York,* 1974: Alfred A. Knopf, N.Y.
Chamber of Commerce, Borough of Queens. *Queens Borough, New York City, 1910-1920: The Borough of Homes and Industry* (Walter I. Willis, ed.), 1920, Chamber of Commerce, Queens, N.Y.
Chamber of Commerce of the Borough of Queens. *"Queensboro" Monthly Magazine,* 1914-Present, issued annually.
Chapman Publishing Co. *Portrait and Biographical Record of Queens County, Long Island, N.Y.* Published 1896.
Fischler, Stan. *Uptown, Downtown. A History of the New York City Subways;* N.Y. 1976, Hawthorn Books.
Hazelton, Henry I. *The Boroughs of Brooklyn and Queens.* 3 vols. New York: Lewis Historical Publishing Company, 1925.
Kelley, F.B. *Historical Guide to the City of New York.* New York: Frederick A. Stokes and Company, 1909.
Kelsey, J.S. *History of Long Island City.* Long Island City: Long Island Star Publishing Company, 1896.
Lawson, Harriet Dondero. *Old Flushing;* pub. Dondero Lawson Co. N.Y. 1952.
Mandeville, G. Henry. *Flushing Past and Present.* J. Egbert, Printer, 1860.
Meigs, Alice H. *Description of Private and Family Cemeteries in the Borough of Queens.* Jamaica: Queens Borough Public Library, 1932.
Munsell and Company. *History of Queens County, N.Y.* New York: W.W. Munsell and Company, 1882.
Old Woodhaven: A Victorian Village. Flushing: Queens Historical Society, 1978.
Onderdonk. J. Jr. *Queens County in Olden Times.* Jamaica, 1865.
Onderdonk, Henry J. *Revolutionary Incidents of Queens County.* New York: Leavitt, Trow and Company, 1846.

Peyer, Jean. *Jamaica, Long Island, 1656-1776: A Study of the Roots of American Urbanism.* Ph.D. thesis, University Microfilms, International, Ann Arbor, 1978.
Peyer, Jean. "*Class Structure and Social Mobility in Jamaica, Queens.*" An article in the *Journal of Long Island History,* 14:1, PP 34-47.
Riker, James R., Jr. *Annals of Newtown in Queens County,* N.Y. New York: D. Fanshaw and Company, 1852.
Seyfried, Vincent F. *A Comprehensive History of the Long Island Rail Road.* 7 vols. Garden City: Salisbury Printers, 1960-1982.
Seyfried, Vincent F. *A History of Queens Village.*
Skal, George von. *Illustrated History of the Borough of Queens.* New York: F.T. Smiley Publishing Company, 1908.
Stankowski, Barbara W. Maspeth, *Our Town.* Maspeth: Federal Savings and Loan Association, 1978.
Thompson, Benjamin F. *History of Long Island.* 3 vol. 1918: Robert H. Dodd Co., N.Y.
Valles, Madeline B. *History of Flushing.* Master's Thesis, Columbia, 1938.
Von Skal, George. *Illustrated History of the Borough of Queens,* New York City. 1908: F.T. Smiley Publ. Co.
Waller, Henry. *History of the Town of Flushing, Long Island, New York.* 1899: J.H. Ridenour, Flushing.

page 29—"Aunt Sue's Stories" copyright 1926 by Alfred A. Knopf, Inc. and renewed 1954 by Langston Hughes. Reprinted from SELECTED POEMS OF LANGSTON HUGHES, by Langston Hughes, by permission of Alfred A. Knopf, Inc.
page 30 (top)—Wichita State University Endowment Association Art Collection. Edwin A. Ulrich Museum of Art, Wichita Kansas.
page 30 (bottom)—Museum of Art, Carnegie Institute, Pittsburgh; Museum Purchase: Patrons Art Fund, 1939.
page 31—Photo, courtesy of Nancy Aaron.
page 33—Photo, courtesy of Nancy Aaron.
page 34 (top left)—Photo, courtesy of Queens Chamber of Commerce.
page 34 (bottom left)—Figure, courtesy of Queens Chamber of Commerce.
page 34 (bottom right)—Oral history, courtesy of Nancy Aaron.
page 35—"Mill Doors" from CHICAGO POEMS by Carl Sandburg, copyright 1916 by Holt, Rinehart and Winston, Inc.; copyright 1944 by Carl Sandburg. Reprinted by permission of Harcourt Brace Jovanovich, Inc.
page 36—Courtesy of Terry Dintenfass Gallery.
pages 36-37—From WE WERE THERE: THE STORY OF WORKING WOMEN IN AMERICA by Barbara Wertheimer. Copyright 1977 by Pantheon Books, a Division of Random House, Inc.
pages 38-39—From THE WOMAN WARRIOR: MEMOIRS OF CHILDHOOD AMONG GHOSTS, by Maxine Hong Kingston. Copyright ©1975, 1976 by Maxine Hong Kingston. Reprinted by permission of Alfred A. Knopf, Inc.
page 41—Photo, courtesy of Nancy Aaron.
page 42—"Sonnet to Negro Soldiers" by Joseph Seaman Cotter Jr.
pages 42–45—© 1946 by Frank Yerby. Reprinted by permission of the William Morris Agency on behalf of the author.
pages 46-47—"Times Square Shoeshine Composition" from JUST GIVE ME A COOL GLASS OF WATER 'FORE I DIE by Maya Angelou. Copyright ©1971 by Maya Angelou. Printed by permission of Random House Inc.
page 48—Photo courtesy of Jess Manyoky.
page 49—Oral history, courtesy of Jess Manyoky.
pages 50-52—"Thanking My Mother for Piano Lessons" by Diane Wakoski. Copyright © 1983 Diane Wakoski.
page 53 (top)—Photo, courtesy of Jess Manyoky
page 53 (bottom)—Figure, courtesy of Queensborough Chamber of Commerce.
page 54—Courtesy of A&P Food Stores.
page 56—Photo, courtesy of Jess Manyoky.
page 59—Photo, courtesy of John Steinway.
page 62—"A Worker Reads History" in SELECTED POEMS, copyright 1947 by Bertolt Brecht and H. R. Hays; renewed 1975 by Stefan S. Brecht and H. R. Hays. Reprinted by permission of Harcourt Brace Jovanovich, Inc.
page 65—Photo, courtesy of Vincent Seyfried.
page 68 (top)—Photo, courtesy of John Steinway.
page 68 (bottom)—Oarsmen at Chatou by Auguste Renoir. National Gallery of Art, Washington. Gift of Sam A. Lewisohn.
page 69—Photo, courtesy of Andrew Malan.
page 71—Photo, courtesy of Philip Sillman.
page 73—Photo, courtesy of Vincent Seyfried.
page 74 (top)—Photo, courtesy of Dora Geipel.
page 74 (bottom)—Courtesy of Vincent Seyfried.
page 76—Photo, courtesy of Dora Geipel.
page 77—Courtesy of Queensborough Public Library.
page 78—Photo, courtesy of Vincent Seyfried.
page 79—Photo, courtesy of Dora Geipel.
page 80—Photo, courtesy of Vincent Seyfried.
page 82—Photo, courtesy of Vincent Seyfried.
page 83—Photo, courtesy of Vincent Seyfried.
page 84—Photo, courtesy of Vincent Seyfried.
page 85—Photo, courtesy of William Kells.
page 86—Photo, courtesy of William Kells.
page 91—Library of Congress.
page 93—A selection from "Big Blonde" from THE PORTABLE DOROTHY PARKER. Copyright 1929 by Dorothy Parker. Copyright © renewed 1957 by Dorothy Parker. Reprinted by permission of Viking Penguin Inc.
page 95—Photo, courtesy of Joseph Duffy.
page 96—Courtesy of Queens Borough President's Office.
page 97—Courtesy, Queensborough Public Library.
page 98 (top)—Copyright 1939 New York News Inc. Reprinted by permission.
page 99—From RABBIT IS RICH by John Updike. Copyright ©1981 by John Updike. Reprinted by permission of Alfred A. Knopf, Inc.

page 105—Photo, courtesy of Eleanor Wallace.
page 106—Photo, courtesy of Dora Geipel.
page 107—Photo, courtesy of Dora Geipel.
page 109 (top)—Photo, courtesy of John Chinkel.
page 109 (bottom)—Photo, courtesy of John Chinkel.
page 110—Photo, courtesy of Frank Gaudett.
page 111 (top)—Photo, courtesy of Dora Geipel.
page 111 (bottom)—Oral history, courtesy of Long Island City resident.
page 112—Courtesy of Queensborough Public Library.
page 116—Courtesy of Queensborough Public Library.
page 118 (top left)—Advertisement, courtesy of Vincent Seyfried.
page 118 (bottom left)—Photo, courtesy of Vincent Seyfried.
page 119—Photo, courtesy of Vincent Seyfried.

page 120 (top)—Photo, courtesy of Vincent Seyfried.
page 120 (bottom)—Photo, courtesy of Queensborough Public Library, Long Island Division, Weber Collection.
page 121—Courtesy of Queens Chamber of Commerce.
page 122—"Incident" from ON THESE I STAND by Countee Cullen. Copyright 1925, by Harper and Row Publishing Inc.; renewed 1953, by Ida M. Cullen, reprinted by permission of the publisher.
page 128—Courtesy of Queensborough Public Library, Long Island Division, Weber Collection.
page 129—Photo, courtesy of Vincent Seyfried.
page 130—Courtesy of Queensborough Public Library, Long Island Division, Weber Collection.
page 131 (bottom)—Museum of the City of New York.
pages 131-133—The lines from "To Brooklyn Bridge" are reprinted from THE BRIDGE by Hart Crane, by permission of Liveright Publishing Corporation. Copyright 1933, ©1958, 1970 by Liveright Publishing Corporation.
page 133 (bottom)—Museum of the City of New York.
page 134—Museum of the City of New York.
page 135—Museum of the City of New York.
page 136—Museum of the City of New York.
page 137—Hirschhorn Museum and Sculpture Garden, Smithsonian Institution.
page 138—From ALEXANDER'S BRIDGE by Willa S. Cather. Copyright 1912, 1922 by Willa Sibert Cather. Copyright renewed, 1950 by Edith Lewis. Reprinted by permission of Houghton Mifflin Company.
page 139—"It is Deep" from HOW I GOT OVAH by Carolyn M. Rodgers. Copyright © 1969 by Carolyn M. Rodgers. Reprinted by permission of Doubleday & Company, Inc.
page 140—Photo, courtesy of Queensborough Public Library, Long Island Division, Weber Collection.
page 142 (left)—Advertisement, courtesy of Long Island Railroad.
page 142 (right)—Advertisement, courtesy of Long Island Railroad.
page 143—Written expressly for the authors by Red Barber.
page 145—Courtesy of Queensborough Public Library.
page 146—Photo, courtesy of Vincent Seyfried.
page 147 (top)—Photo, courtest of Vincent Seyfried.
page 147 (bottom)—Courtesy Queens Chamber of Commerce.
page 148—Courtesy of Mr. and Mrs. Arthur G. Altschul.
page 149—Photo, courtesy of Catharine Heydner.
page 150—Photo, courtesy of Ruthe Mitchell.
page 151 (top)—Photo, courtesy of Nola McNeill.
page 151 (bottom)—Photo, courtesy of Queens Chamber of Commerce.
page 152 (top)—THE GOLDEN AGE OF THE LUXURY CAR ed. G. Hildebrand, 1980, Dover Publications, Inc., N.Y.
page 152 (bottom)—Photo, courtesy of Queens Chamber of Commerce.
pages 153–154—Excerpted from BABBITT by Sinclair Lewis, copyright 1922 by Harcourt Brace Jovanovich, Inc.; renewed 1950 by Sinclair Lewis. Reprinted by permission of the publisher.
page 154—Courtesy of Chrysler Historical Collection.
page 155—Courtesy of Oldsmobile Division, General Motors Corporation.
page 157—Courtesy of Queensborough Public Library, Long Island Division, Weber Collection.
page 159—"Auto Wreck" Copyright 1942 and renewed 1970 by Karl Shapiro. Reprinted from COLLECTED POEMS 1940-1978, by Karl Shapiro by permission of Random House Inc.
page 160 Courtesy of Queensborough Public Library, Long Island Division, Weber Collection.
page 161—Oral history, courtesy of John Abramopaulos.
page 162—Photo, courtesy of John Abramopaulos.
page 163—Charles A. Lindbergh, from *The Spirit of St. Louis* copyright 1953 Charles Scribners Sons; Copyright renewed 1981, Anne Morrow Lindbergh. Reprinted with the permission of Charles Scribners Sons.